DOT PAPER GEOMETRY

WITH OR WITHOUT A GEOBOARD

by Charles Lund

Cuisenaire Company of America, Inc.
12 Church Street, New Rochelle, NY 10801

INTRODUCTION

The activities in this book are designed to be used with a geoboard or with dot paper. Each page is a prelude to more formal activities and exercises in the text-book.

With Geoboards:

The exercises and activities are based on a geoboard with 25 nails in a square array, with some activities using 25-pin geoboards that fit together to form larger rectangles or squares. Anticipating adoption of the metric system, workbook illustrations show the distance between nails or dots in centimeters. For example, the section on length and perimeter assumes that the nails on a geoboard are 4 cm apart. A template for constructing geoboards with these dimensions is provided in the appendix. All of the activities and exercises can be used with non-metric geoboards, with only the answer section for length and perimeter needing to be recomputed.

With Dot Paper:

If geoboards are not available, or if students do not need work with concrete models, then the dot paper activities can be completed directly on the worksheets. Many teachers use the dot paper approach as a valuable follow-up to the earlier geoboard experience. In order to use this book with and without a geoboard, answers for both approaches are provided in the appendix.

Applications and Use:

Dot Paper Geometry is intended for students in grades 4-8. The book has been successfully field tested with large and small groups and with all ability levels. While a wide variety of topics are presented in the book, a strong emphasis is placed on geometry and measurement.

Each section of the book has been sequenced developmentally, and provides teachers with a source of "warm-up lessons" to use before more formal treat-ment of a topic in the standard textbook. The philosophical framework used throughout the book is that children need experiences with concrete and pic-torial work prior to the usual textbook lesson.

The appendix at the back of the book contains selected answers and brief com-ments. There are also several pages of black-line masters to produce record sheets, protractors, and rulers. The appendix includes review tests which teachers may choose to use at different stages to check student progress through this book.

It is hoped the activities in Dot Paper Geometry give teachers a valuable source of motivating concrete or pictorial lessons to help students become more ac-tively involved with the learning of new mathematical concepts.

DOT PAPER GEOMETRY
CUISENAIRE

Table of Contents

Metric Length and Perimeter of Polygons

Angle Measure

More Prediction Problems

Masters for Teachers

Quizzes

Answers and Commentary

Geoboard Letter Polygons

For each of the letters below, count the number of nails <u>touched</u> by the rubber band. Count the number of sides. The letters F and G are left for you to make.

EXAMPLE

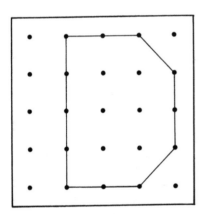

A. Number of nails: __14__
    Number of sides: __10__

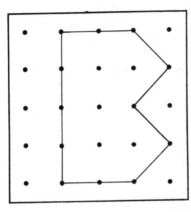

B. Number of nails: _____
    Number of sides: _____

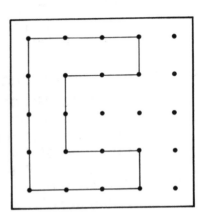

C. Number of nails: _____
    Number of sides: _____

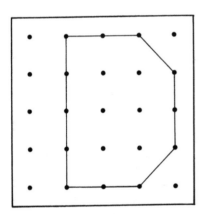

D. Number of nails: _____
    Number of sides: _____

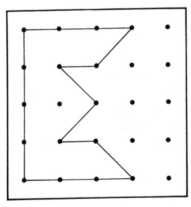

E. Number of nails: _____
    Number of sides: _____

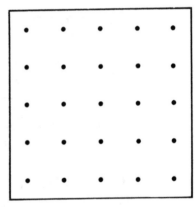

F. Number of nails: _____
    Number of sides: _____

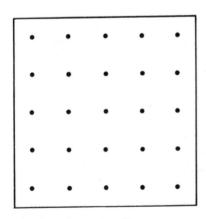

G. Number of nails: _____
    Number of sides: _____

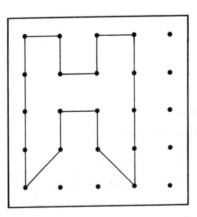

H. Number of nails: _____
    Number of sides: _____

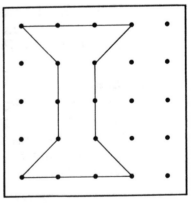

I. Number of nails: _____
    Number of sides: _____

Geoboard Numeral Polygons

For each of the numerals below, count the number of nails <u>touched</u> by the rubber band. Count the number of sides. The numerals 6 and 7 are left for you to make.

EXAMPLE

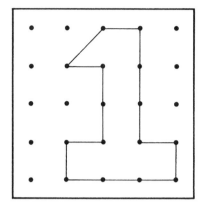

1. Number of nails: <u>15</u>
    Number of sides: <u>10</u>

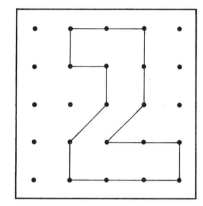

2. Number of nails: _____
    Number of sides: _____

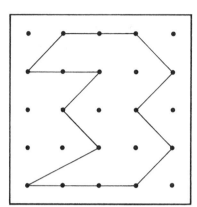

3. Number of nails: _____
    Number of sides: _____

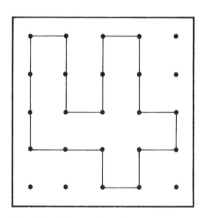

4. Number of nails: _____
    Number of sides: _____

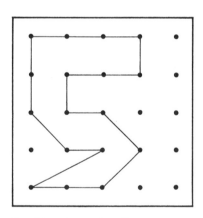

5. Number of nails: _____
    Number of sides: _____

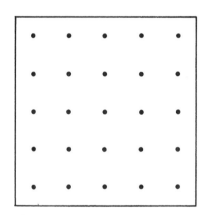

6. Number of nails: _____
    Number of sides: _____

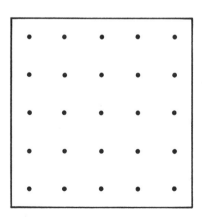

7. Number of nails: _____
    Number of sides: _____

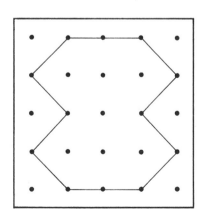

8. Number of nails: _____
    Number of sides: _____

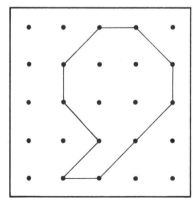

9. Number of nails: _____
    Number of sides: _____

Parallel Lines

For each of the given segments, find three segments parallel to it. Show your answers on the same geoboard. Compare your answers with your classmates.

EXAMPLE

1.

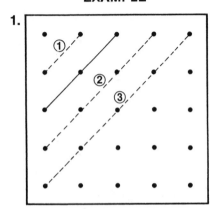

Can you find three more?

2.

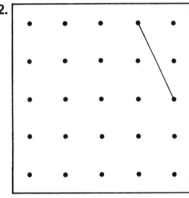

3.

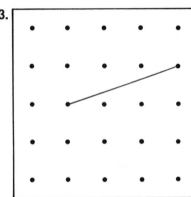

4.

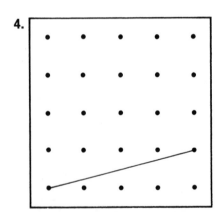

5.

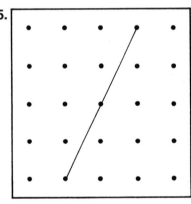

6.

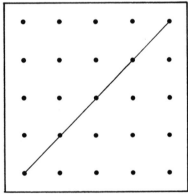

7.

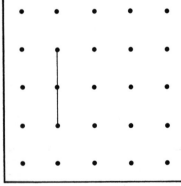

8.

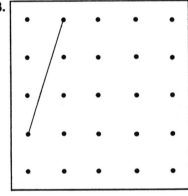

9.

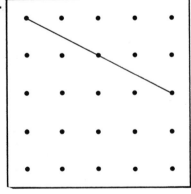

Perpendicular Lines

For each of the given segments, find three segments perpendicular to it. Show your answers on the same geoboard. Compare your answers with your classmates.

EXAMPLE

1.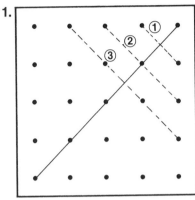

Can you find three more?

2.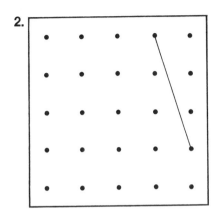

3.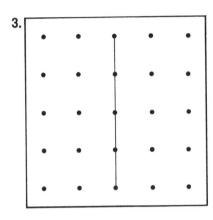

4.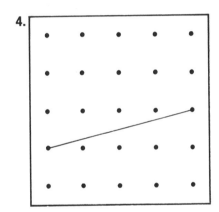

5.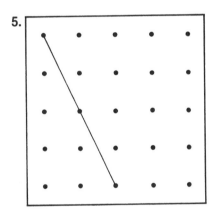

6.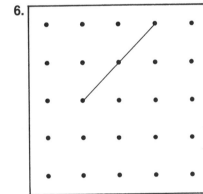

7.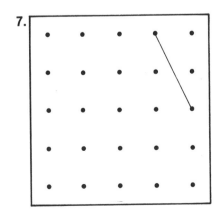

8.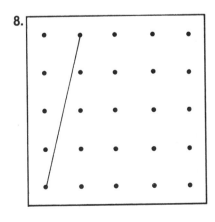

9.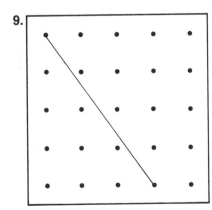

Constructing Polygons

Make a figure which fits the given conditions. Compare your polygons with your classmates. How are they alike and different?

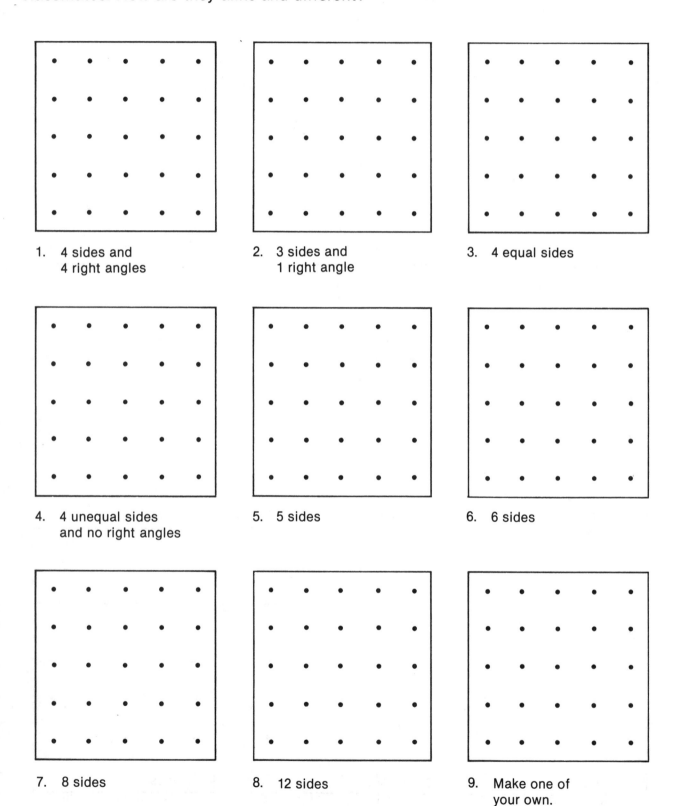

1. 4 sides and
    4 right angles

2. 3 sides and
    1 right angle

3. 4 equal sides

4. 4 unequal sides
    and no right angles

5. 5 sides

6. 6 sides

7. 8 sides

8. 12 sides

9. Make one of
    your own.

Building Polygons with Parallel and Perpendicular Sides

Make a figure which fits the given conditions. Compare your polygons with your classmates. How are they alike and different?

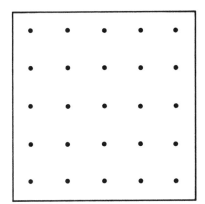

1. 4 sides —
   exactly 2 sides parallel

2. 4 sides —
   2 pairs of sides parallel

3. 4 sides —
   exactly 1 pair of sides
   perpendicular

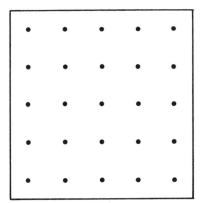

4. 4 sides —
   2 pairs of sides
   perpendicular

5. 4 sides —
   no sides parallel

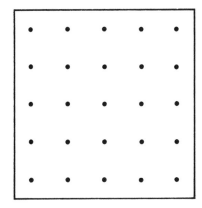

6. 4 sides —
   no sides perpendicular

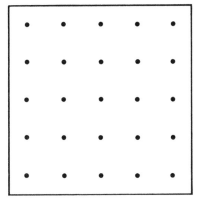

7. 3 sides —
   2 sides perpendicular

8. 5 sides —
   exactly 1 pair of parallel
   sides

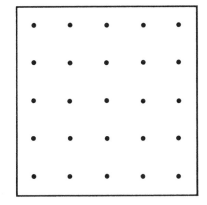

9. 6 sides —
   3 pairs of parallel sides

Building Polygons with Given Lengths

Make a polygon with the given lengths as sides. Sketch your polygons on the grids provided. Compare your answers with your classmates.

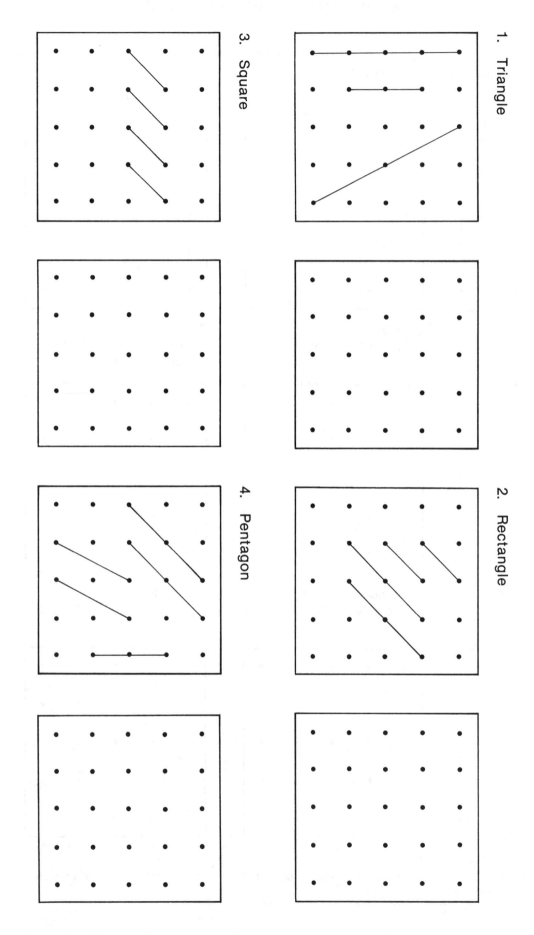

1. Triangle

2. Rectangle

3. Square

4. Pentagon

Building Polygons with Given Lengths

Make a polygon with the given lengths as sides. Sketch your polygons on the grids provided. Compare your answers with your classmates.

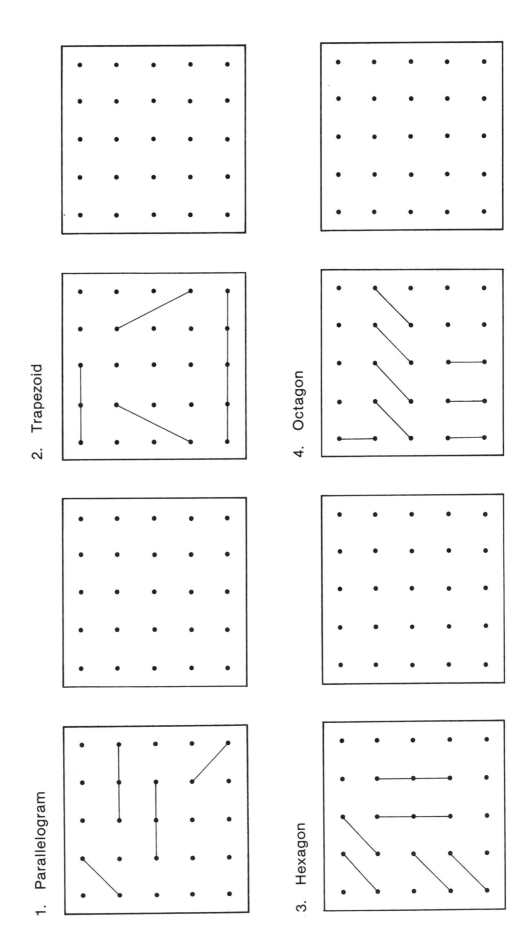

1. Parallelogram

2. Trapezoid

3. Hexagon

4. Octagon

Tic Tac Toe Using Ordered Pairs of Whole Numbers

Many games involve the use of numbers and a grid. Shown below is a move-by-move account of a game that is similar to Tic Tac Toe. Study the moves to see if you can figure out how to play.

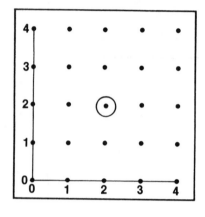

Chuck's
first move: (2,2)

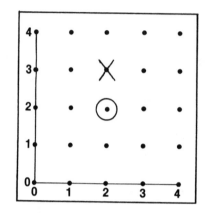

Mary's
first move: (2,3)

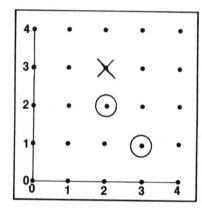

Chuck's
second move: (3,1)

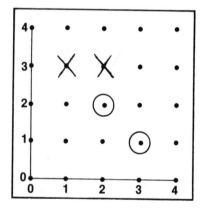

Mary's
second move: (1,3)

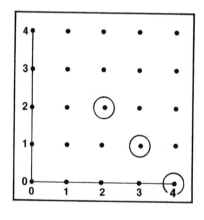

Chuck's
third move, (4,0) and he wins already!

3 in a row, column,
or diagonal wins.

Play at least three games of Tic Tac Toe using the grids on the next page.

Playing Tic Tac Toe

Choose a classmate and play three games of Tic Tac Toe as described on the previous page. Record the ordered pairs of numbers for each move. In order to win, a player must get 3 markers in a row, column or diagonal.

GAME 1
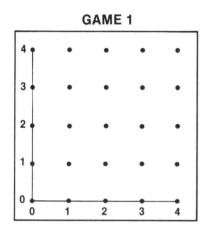

GAME 2
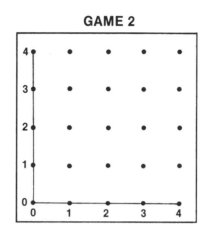

GAME 3
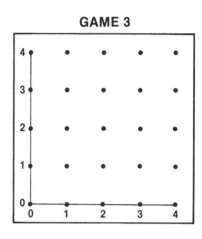

Ordered pairs:

Player 1	Player 2
(,)	(,)
(,)	(,)
(,)	(,)
(,)	(,)

Ordered pairs:

Player 1	Player 2
(,)	(,)
(,)	(,)
(,)	(,)
(,)	(,)

Ordered pairs:

Player 1	Player 2
(,)	(,)
(,)	(,)
(,)	(,)
(,)	(,)

Now modify the rules so that it takes 4 in a row, column, or diagonal to win.

GAME 4
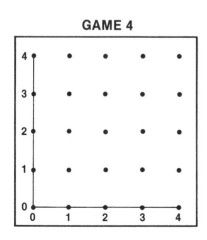

GAME 5
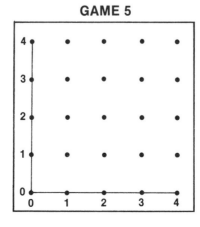

GAME 6
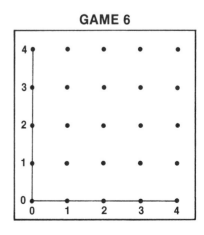

Record your ordered pairs for the three games of 4-in-a row on a sheet of paper.

Plotting Ordered Pairs

1. What letter of the alphabet does this set of ordered pairs form? _____

$$\left\{ \begin{array}{llll} (0,4) & (1,4) & (2,4) & (3,4) \\ (0,3) & (0,2) & (0,1) & (0,0) \\ (1,0) & (2,0) & (3,0) & \end{array} \right\}$$

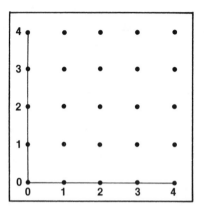

2. What letter of the alphabet does this set of ordered pairs form? _____

$$\left\{ \begin{array}{llll} (0,0) & (0,1) & (0,2) & (0,3) \\ (0,4) & (1,4) & (1,0) & (1,2) \\ (2,0) & (2,4) & (2,2) & \end{array} \right\}$$

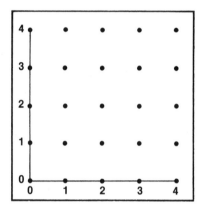

Each of these sets of ordered pairs forms the outline of a familiar geometric shape. Name each one.

Make one of your own

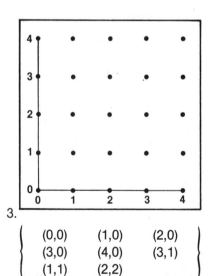

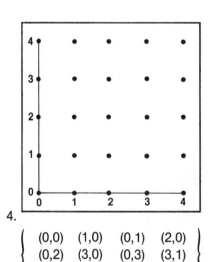

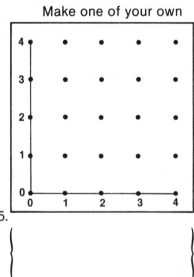

3.
$$\left\{ \begin{array}{lll} (0,0) & (1,0) & (2,0) \\ (3,0) & (4,0) & (3,1) \\ (1,1) & (2,2) & \end{array} \right\}$$

Shape: _____

4.
$$\left\{ \begin{array}{llll} (0,0) & (1,0) & (0,1) & (2,0) \\ (0,2) & (3,0) & (0,3) & (3,1) \\ (3,2) & (3,3) & (2,3) & (1,3) \end{array} \right\}$$

Shape: _____

5.
$$\left(\right)$$

Shape: _____

Writing Ordered Pairs

Make the sets of ordered pairs to form these letters of the alphabet, P A L.

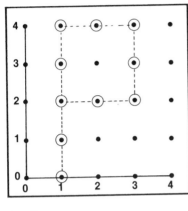

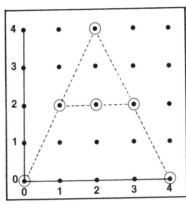

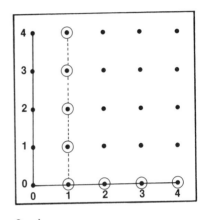

1. P:

2. A:

3. L:

$\left\{ \rule{0pt}{4em} \right.$ $\left\{ \rule{0pt}{4em} \right.$ $\left\{ \rule{0pt}{4em} \right.$ $\left\{ \rule{0pt}{4em} \right.$

Write the ordered pairs used to form these numerals.

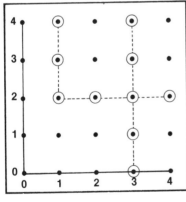

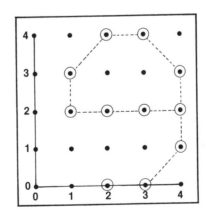

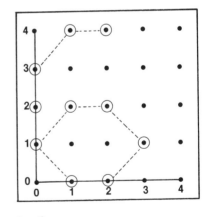

4. 4:

5. 9:

6. 6:

$\left\{ \rule{0pt}{4em} \right.$ $\left\{ \rule{0pt}{4em} \right.$ $\left\{ \rule{0pt}{4em} \right.$ $\left\{ \rule{0pt}{4em} \right.$

DOT PAPER GEOMETRY © 1980 Cuisenaire Co. of America, Inc.

Classifying Triangles by Their Sides

Triangles are sometimes classified according to the lengths of their sides.

Scalene triangle: no two sides have the same length.
Isosceles triangle: two sides have the same length
Equilateral triangle: all three sides have the same length

Plot the ordered pairs for the vertices (corners) of each triangle below. Classify each triangle by circling the category to which it belongs. Use a metric ruler to help when necessary.

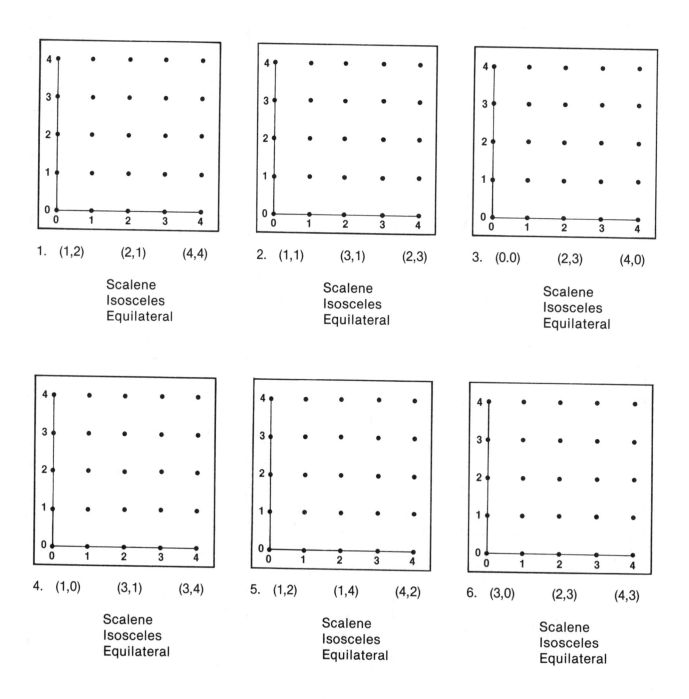

1. (1,2)   (2,1)   (4,4)

Scalene
Isosceles
Equilateral

2. (1,1)   (3,1)   (2,3)

Scalene
Isosceles
Equilateral

3. (0.0)   (2,3)   (4,0)

Scalene
Isosceles
Equilateral

4. (1,0)   (3,1)   (3,4)

Scalene
Isosceles
Equilateral

5. (1,2)   (1,4)   (4,2)

Scalene
Isosceles
Equilateral

6. (3,0)   (2,3)   (4,3)

Scalene
Isosceles
Equilateral

Classifying Triangles by Their Angles

Triangles are also sometimes classified according to the measurement of their angles.

Acute triangle: all 3 angles have a measure less than 90°.
Obtuse triangle: one angle measures more than 90°.
Right triangle: one angle measures 90°.

Plot the ordered pairs for the vertices (corners) of each triangle below. Classify each triangle by circling the category to which it belongs. Use a protractor to help when necessary.

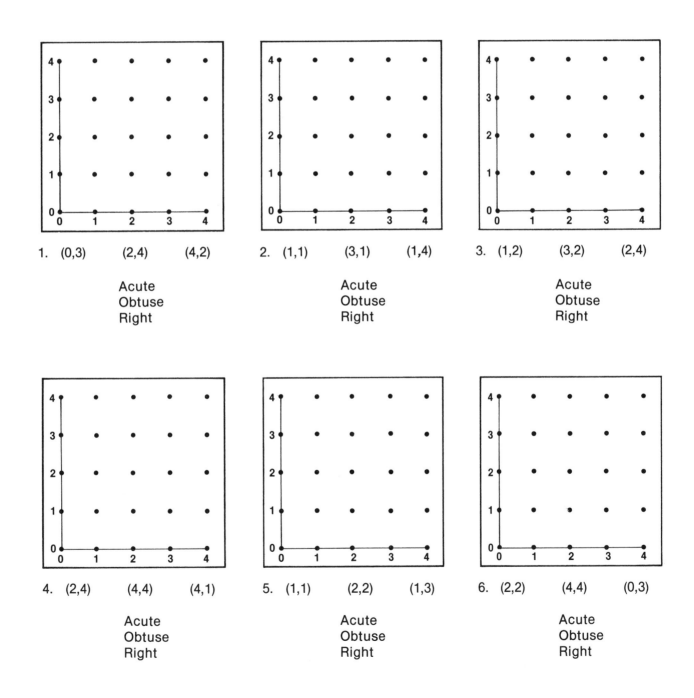

1. (0,3)   (2,4)   (4,2)

Acute
Obtuse
Right

2. (1,1)   (3,1)   (1,4)

Acute
Obtuse
Right

3. (1,2)   (3,2)   (2,4)

Acute
Obtuse
Right

4. (2,4)   (4,4)   (4,1)

Acute
Obtuse
Right

5. (1,1)   (2,2)   (1,3)

Acute
Obtuse
Right

6. (2,2)   (4,4)   (0,3)

Acute
Obtuse
Right

Classifying Quadrilaterals

Plot the ordered pairs for the vertices (corners) of each quadrilateral below. Classify each figure by circling all of the categories to which it belongs. Use a metric ruler and protractor to help when necessary.

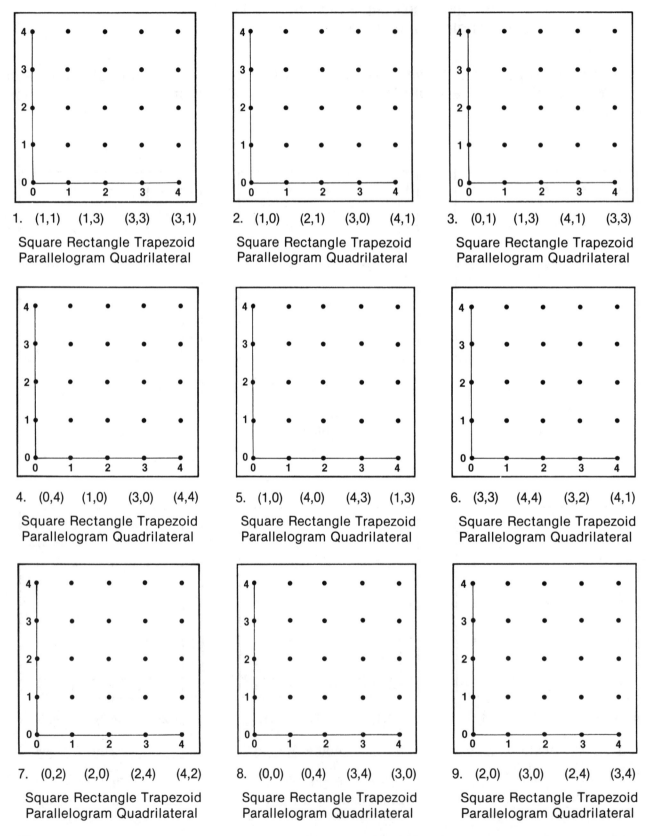

1. (1,1) (1,3) (3,3) (3,1)
Square Rectangle Trapezoid
Parallelogram Quadrilateral

2. (1,0) (2,1) (3,0) (4,1)
Square Rectangle Trapezoid
Parallelogram Quadrilateral

3. (0,1) (1,3) (4,1) (3,3)
Square Rectangle Trapezoid
Parallelogram Quadrilateral

4. (0,4) (1,0) (3,0) (4,4)
Square Rectangle Trapezoid
Parallelogram Quadrilateral

5. (1,0) (4,0) (4,3) (1,3)
Square Rectangle Trapezoid
Parallelogram Quadrilateral

6. (3,3) (4,4) (3,2) (4,1)
Square Rectangle Trapezoid
Parallelogram Quadrilateral

7. (0,2) (2,0) (2,4) (4,2)
Square Rectangle Trapezoid
Parallelogram Quadrilateral

8. (0,0) (0,4) (3,4) (3,0)
Square Rectangle Trapezoid
Parallelogram Quadrilateral

9. (2,0) (3,0) (2,4) (3,4)
Square Rectangle Trapezoid
Parallelogram Quadrilateral

DOT PAPER GEOMETRY © 1980 Cuisenaire Co. of America, Inc.

Classifying Quadrilaterals

Plot the ordered pairs for the vertices (corners) of each quadrilateral below. Classify each figure by circling all of the categories to which it belongs. Use a metric ruler and protractor to help when necessary.

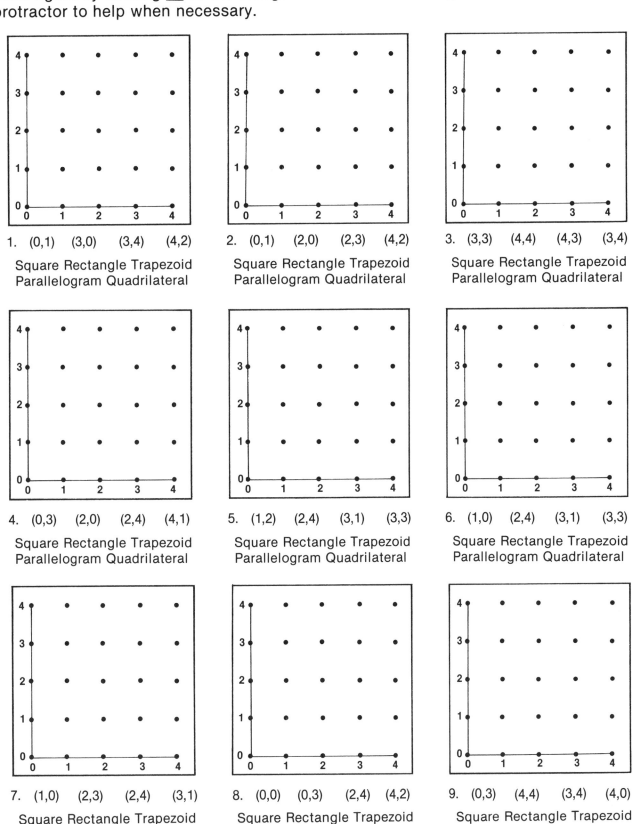

1. (0,1) (3,0) (3,4) (4,2)
    Square Rectangle Trapezoid
    Parallelogram Quadrilateral

2. (0,1) (2,0) (2,3) (4,2)
    Square Rectangle Trapezoid
    Parallelogram Quadrilateral

3. (3,3) (4,4) (4,3) (3,4)
    Square Rectangle Trapezoid
    Parallelogram Quadrilateral

4. (0,3) (2,0) (2,4) (4,1)
    Square Rectangle Trapezoid
    Parallelogram Quadrilateral

5. (1,2) (2,4) (3,1) (3,3)
    Square Rectangle Trapezoid
    Parallelogram Quadrilateral

6. (1,0) (2,4) (3,1) (3,3)
    Square Rectangle Trapezoid
    Parallelogram Quadrilateral

7. (1,0) (2,3) (2,4) (3,1)
    Square Rectangle Trapezoid
    Parallelogram Quadrilateral

8. (0,0) (0,3) (2,4) (4,2)
    Square Rectangle Trapezoid
    Parallelogram Quadrilateral

9. (0,3) (4,4) (3,4) (4,0)
    Square Rectangle Trapezoid
    Parallelogram Quadrilateral

Congruent Polygons

Polygons that are the same size and shape are <u>congruent</u>. They do not need to have the same position or orientation. Here are three congruent pentagons. If you wish, cut them out to check that they match.

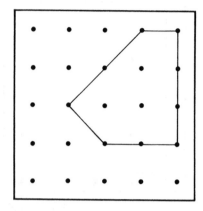 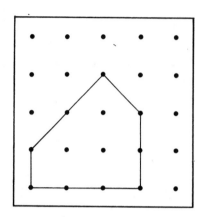 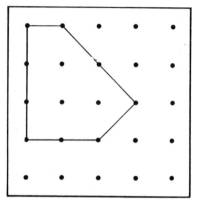

For each of the given figures, sketch two figures congruent to it on the grids provided.

1.

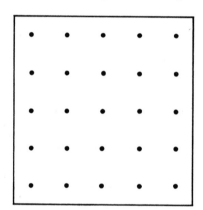

2.

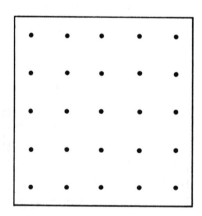 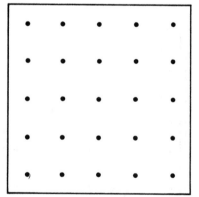

Symmetry

A figure has a <u>line of symmetry</u> if it can be folded so that the two parts are congruent. Examples of three figures with line symmetry are shown below.

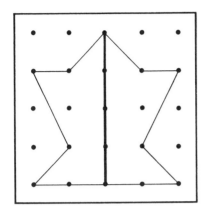

Complete each figure below so that the given line is a line of symmetry.

1.
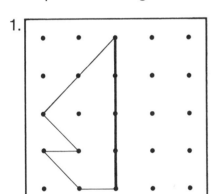

2.
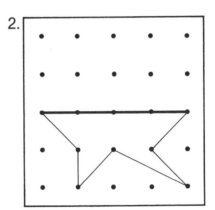

3.
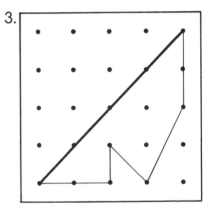

Draw both parts for these figures so that the given line is a line of symmetry. Compare your answers with your classmates.

4.
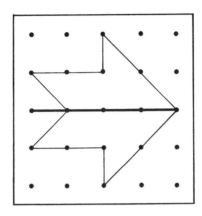

5.
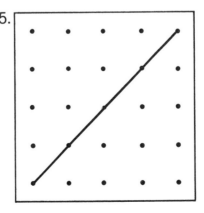

6.
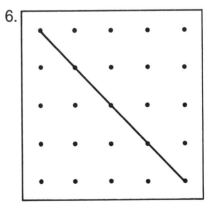

Dividing a Geoboard into Equal Parts

1. Find two more ways of dividing a geoboard into halves. Sketch your answers on the grids provided below.

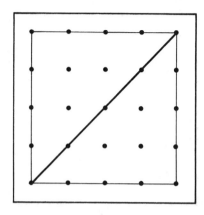 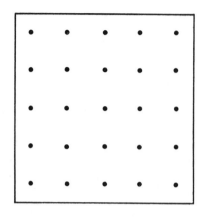 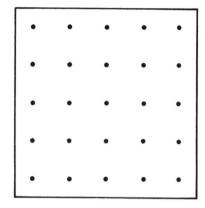

2. Find three different ways of dividing a geoboard into fourths. Sketch your answers on the grids provided below.

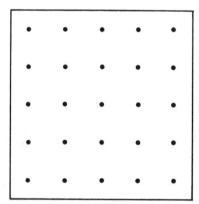 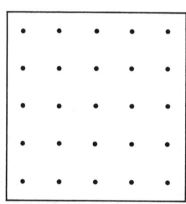 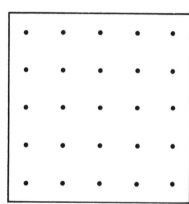

3. Find three different ways of dividing a geoboard into eighths. Sketch your answers on the grids provided below.

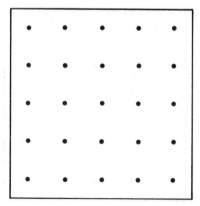 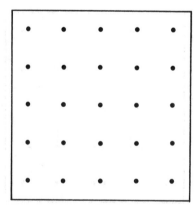 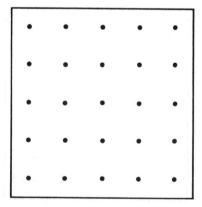

Building and Shading Fractions

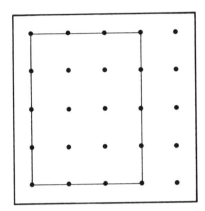

1. Divide into sixths.
   Shade 5/6.

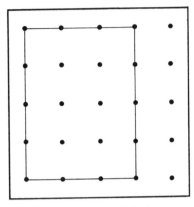

2. Divide into thirds.
   Shade 2/3.

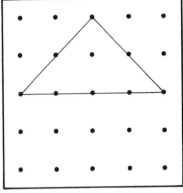

3. Divide into halves.
   Shade 1/2.

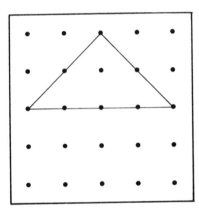

4. Divide into fourths.
   Shade 1/4.

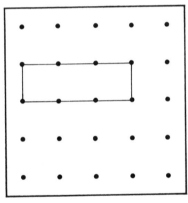

5. Divide into sixths.
   Shade 2/6.

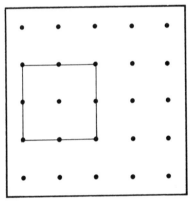

6. Divide into eighths.
   Shade 3/8.

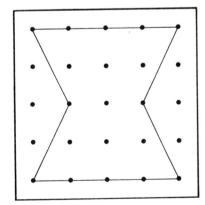

7. Divide into fourths.
   Shade 3/4.

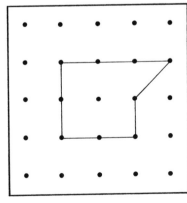

8. Divide into thirds.
   Shade 1/3.

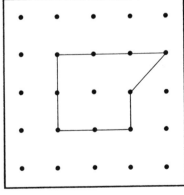

9. Divide into ninths.
   Shade 3/9.

DOT PAPER GEOMETRY © 1980 Cuisenaire Co. of America, Inc.

Equivalent Fractions

1. Suppose ABCD = 1:

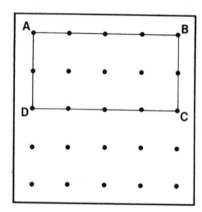

Build and sketch 1/2.

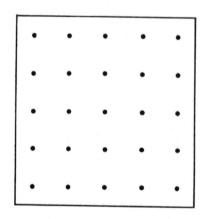

Build and sketch 2/4.

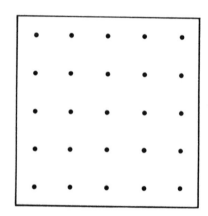

2. Suppose EFGH = 1:

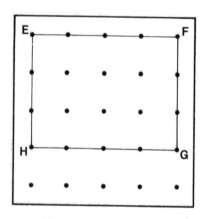

Build and sketch 1/3.

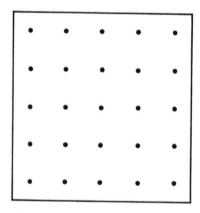

Build and sketch 2/6.

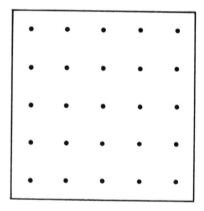

3. Suppose IJKL = 1:

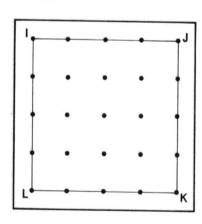

Build and sketch 3/4.

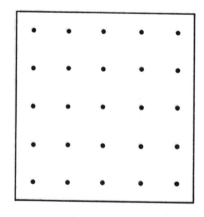

Build and sketch 6/8.

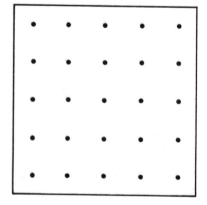

Equivalent Fractions

1. Suppose MNO = 1:

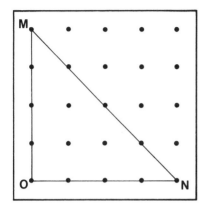

Build and sketch 1/4.

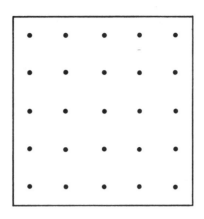

Build and sketch 2/8.

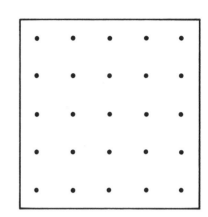

2. Suppose AEMJ = 1:

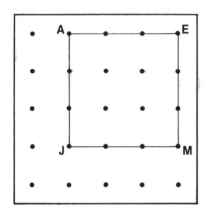

Build and sketch 2/3.

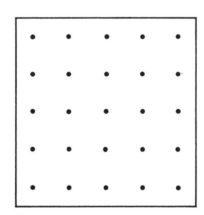

Build and sketch 4/6.

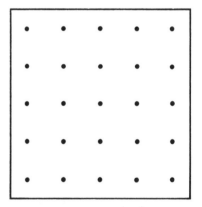

3. Suppose BFNRK = 1:

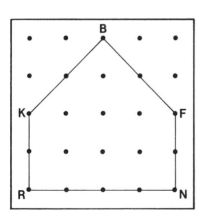

Build and sketch 1/6.

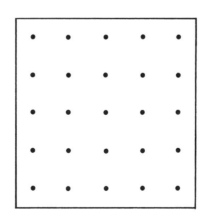

Build and sketch 2/12.

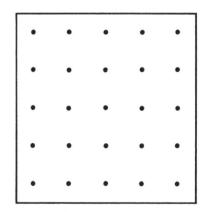

Addition of Fractions

Chris and Chuck made up a mystery fractions activity.

They started with rectangle ABDC = 1,

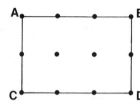

Then each person secretly built a fraction.

Chris built (1/2) on his geoboard:

Chuck built (1/3) on his geoboard:

Next they put their two fractions

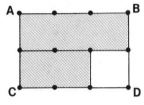

together and recorded their answer.

$$\frac{1}{2} + \frac{1}{3} = \frac{5}{6}$$

Get a partner and try Chris and Chuck's activity. For each problem start with the given unit. Make up a problem and then record your problem and answer. Make up 3 problems for each given unit.

Given Unit	Addition Problems and Answers

1. Start with EFHG = 1

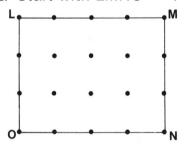

1. a.

    b.

    c.

2. Start with LMNO = 1

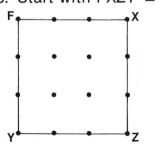

2. a.

    b.

    c.

3. Start with FXZY = 1

3. a.

    b.

    c.

Subtraction of Fractions

Shelly and Diane modified Chris and Chuck's mystery fractions activity for use with subtraction.

They started with ABCD = 1

Then each person secretly built a fraction.   Shelly built $\frac{5}{6}$:

Diane built $\frac{1}{6}$:

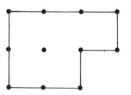

Next they found the difference of their 2 fractions and recorded their answer.

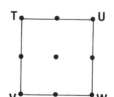

$$\frac{5}{6} - \frac{1}{6} = \frac{4}{6} \text{ or } \frac{2}{3}$$

Get a partner and try Shelly and Diane's activity. For each problem, start with the given unit. Make up a problem and then find and record the answer. Build three problems for each given unit.

<u>**Given Unit**</u>   **Subtraction Problems and Answers**

1. Start with TUWV = 1

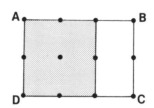

1. a.

   b.

   c.

2. Start with NOQP = 1

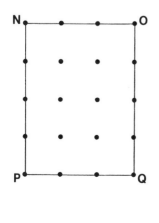

2. a.

   b.

   c.

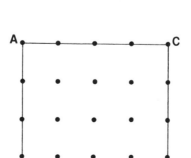

3. a.

   b.

3. Start with ACMX = 1   c.

Multiplication of Fractions

Read through this discussion and then complete the exercises which follow.

Problem: What is $\frac{1}{4}$ x $\frac{1}{3}$ = _____

Step 3. Complete a rectangle using DA and AB as two of the sides.

Step 1. Suppose AB = 1

Step 2. Suppose DA = 1

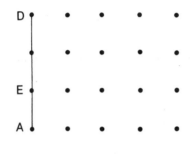

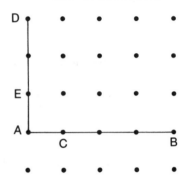

Then how much is AC? _____

Then how much is EA? _____

Step 4. Build a small rectangle using EA and AC as sides inside your rectangle.

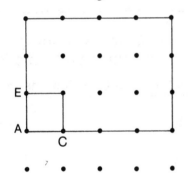

Step 5. If  = 1, what is the area of the rectangle you built in Step 3?

12

Step 6. What fraction of the total rectangle is your small rectangle?

$\frac{1}{12}$

Step 7.

$\frac{1}{4}$ x $\frac{1}{3}$ = $\frac{1}{12}$

Use this sample type of procedure to multiply these fractions.

1.

$\frac{1}{2}$ x $\frac{1}{3}$ = _____

2.

$\frac{1}{4}$ x $\frac{2}{3}$ = _____

3.

$\frac{3}{4}$ x $\frac{2}{3}$ = _____

Division of Fractions

1. Suppose ABDC = 1   Then show:

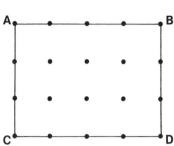

a. $\frac{1}{3}$ b. $\frac{1}{4}$

c. $\frac{1}{2}$ d. $\frac{1}{6}$ e. $\frac{1}{12}$

Problems 2 – 12: Use the above diagrams to help you solve these problems.

2. How many $\frac{1}{3}$s in 1? _____ $1 \div \frac{1}{3} =$ _____

3. How many $\frac{1}{2}$s in 1? _____ $1 \div \frac{1}{2}$ _____

4. How many $\frac{1}{4}$s in 1? _____ $1 \div \frac{1}{4} =$ _____

5. How many $\frac{1}{6}$s in 1? _____ $1 \div \frac{1}{6} =$ _____

6. How many $\frac{1}{12}$s in 1? _____ $1 \div \frac{1}{12} =$ _____

7. How many $\frac{1}{4}$s in 2? _____ $2 \div \frac{1}{4} =$ _____

8. How many $\frac{1}{4}$s in $\frac{1}{2}$? _____ $\frac{1}{2} \div \frac{1}{4} =$ _____

9. How many $\frac{1}{12}$s in $\frac{1}{3}$? _____ $\frac{1}{3} \div \frac{1}{12} =$ _____

10. How many $\frac{1}{6}$s in $\frac{1}{3}$? _____ $\frac{1}{3} \div \frac{1}{6} =$ _____

11. How many $\frac{1}{3}$s in 4? _____ $4 \div \frac{1}{3} =$ _____

12. How many $\frac{1}{6}$s in 3? _____ $3 \div \frac{1}{6} =$ _____

DOT PAPER GEOMETRY © 1980 Cuisenaire Co. of America, Inc.

Naming Percents

Percent means hundredths. Pretend that 4 geoboards are placed together so that there are 100 nails.

Example: Count the number of nails that have been circled.
There are 17 out of the 100.

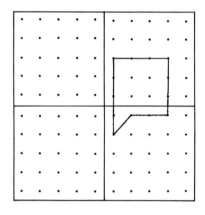

Fraction	Decimal	Percent
$\frac{17}{100}$	.17	17%

Count the number of nails that have been circled. Express the amount as a fraction, decimal, and percent.

	Fraction	Decimal	Percent

1.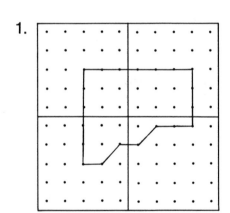

_____   _____   _____

2.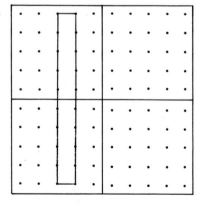

_____   _____   _____

Naming Percents

Count the number of nails that have been circled. Express that amount as a fraction, decimal, and percent.

	Fraction	Decimal	Percent

More Percent

From your previous work with percent on a geoboard you know that percent means hundredths. In this lesson you will practice converting from a fraction whose denominator is not 100 to a percent. Study problem 1 to help you get started.

	Fraction of Nails Enclosed	Equivalent Fraction With Denominator of One Hundred	Percent

1.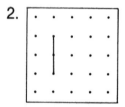

$$\frac{8}{25} \times \left(\frac{4}{4}\right) \quad = \quad \frac{32}{100} \quad = \quad \underline{32\%}$$

2.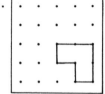

$$\frac{}{25} \qquad\qquad \frac{}{100} \qquad\qquad \underline{}$$

3.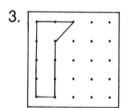

$$\frac{}{25} \qquad\qquad \frac{}{100} \qquad\qquad \underline{}$$

4.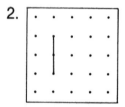

$$\frac{}{25} \qquad\qquad \frac{}{100} \qquad\qquad \underline{}$$

5.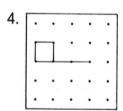

$$\frac{}{25} \qquad\qquad \frac{}{100} \qquad\qquad \underline{}$$

6.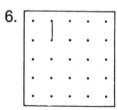

$$\frac{}{25} \qquad\qquad \frac{}{100} \qquad\qquad \underline{}$$

DOT PAPER GEOMETRY © 1980 Cuisenaire Co. of America, Inc.

More Percent

This time you are given the precent. Work backwards to write a fraction with denominator 100. Then write the equivalent fraction with denominator 25. Finally enclose the correct number of nails on the geoboard.

	Fraction of Nails Enclosed	Fraction with Denominator of One Hundred	Percent
1.	$\dfrac{5}{25}$	$\dfrac{20}{100}$	20%
2.	$\dfrac{}{25}$	$\dfrac{}{100}$	60%
3.	$\dfrac{}{25}$	$\dfrac{}{100}$	48%
4.	$\dfrac{}{25}$	$\dfrac{}{100}$	88%
5.	$\dfrac{}{25}$	$\dfrac{}{100}$	40%
6.	$\dfrac{}{25}$	$\dfrac{}{100}$	12%

Area of Rectangles

Suppose 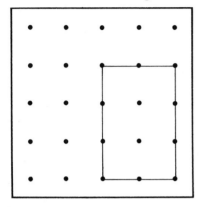 has an area of 1 square unit. Find the area of each of these rectangles. Mark an X in rectangles that are also squares.

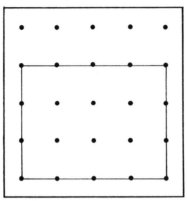

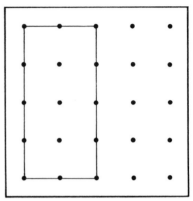

1. Area = _____

2. Area = _____

3. Area = _____

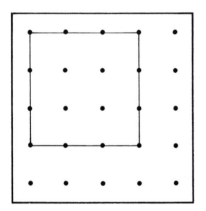

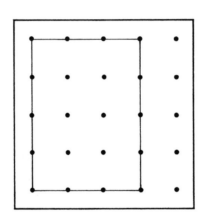

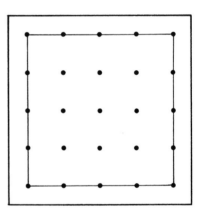

4. Area = _____

5. Area = _____

6. Area = _____

Construct rectangles with the given areas. Compare your rectangles with your classmates.

7. Area = _3_ square units

8. Area = _4_ square units

9. Area = _8_ square units

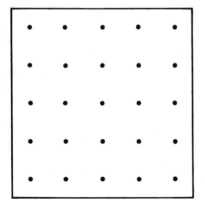

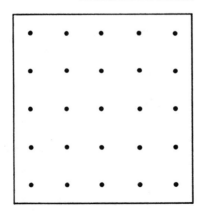

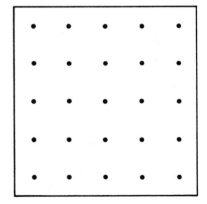

31

Area of Larger Rectangles

Suppose more than one geoboard is placed together as shown. Find the area of each of these larger rectangles. Let 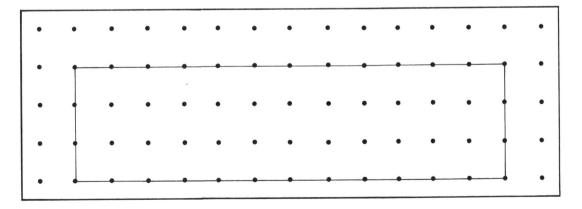 equal 1 square unit of area.

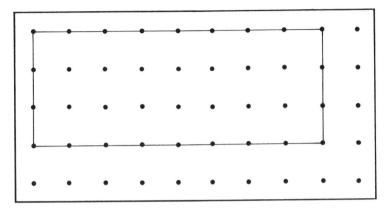

1. Area = _____

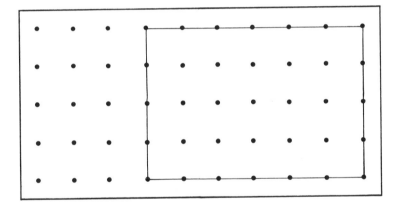

3. Area = _____

2. Area = _____

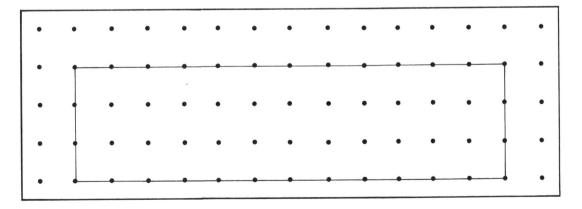

4. Area = _____

Area of Right Triangles

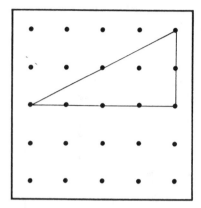

Area = <u>4 square units</u>

One way of finding the area of this right triangle is to look at it as 1/2 of a rectangle with area 8.

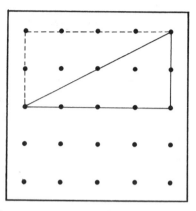

Each right triangle is 1/2 of the rectangle.

Find the area of each right triangle.

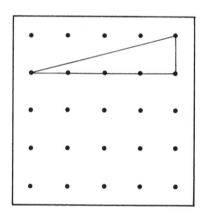

1. Area = _____

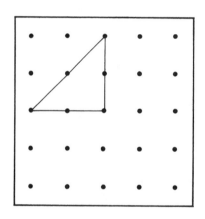

2. Area = _____

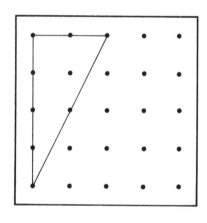

3. Area = _____

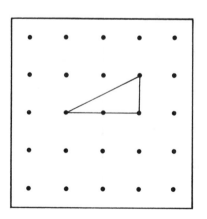

4. Area = _____

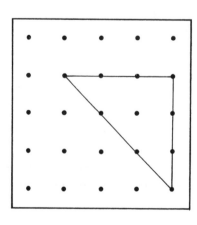

5. Area = _____

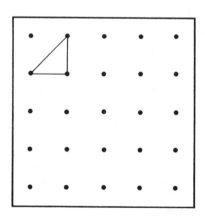

6. Area = _____

Area of Right Triangles

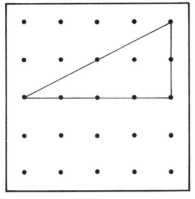

Area = <u>4 square units</u>

Another way of finding the area of this right triangle is to break it up into smaller known areas.

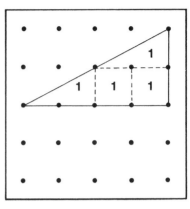

Each small triangle has an area of 1 square unit. The total is 4 square units.

Find the area of each right triangle.

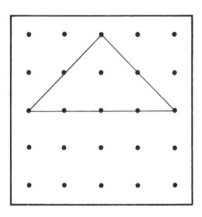

1. Area = _____

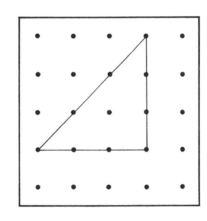

2. Area = _____

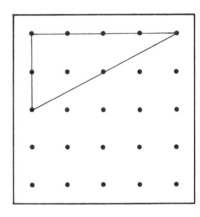

3. Area = _____

Construct right triangles with the given areas. Compare your triangles with your classmates.

4. Area = <u>2 square units</u>

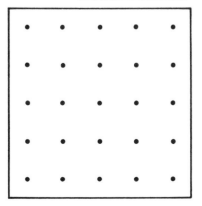

5. Area = <u>6 square units</u>

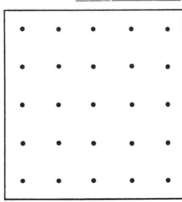

6. Area = <u>8 square units</u>

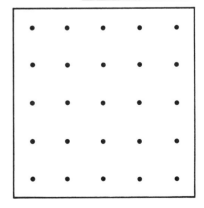

Area of Polygons Using the Chop Strategy

Chris finds the area of polygons using a "chop strategy."

Step 1

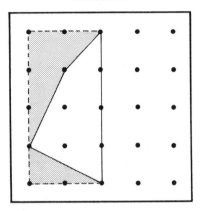

Build a rectangle around the shape. The area of the rectangle is 8. If we can figure out the area of the shaded region and chop it off, we have our answer.

Step 2

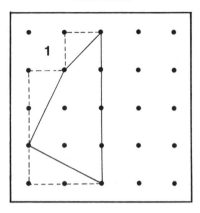

Chop off the square in the upper left corner:

$$8 - 1 = 7$$

Step 3

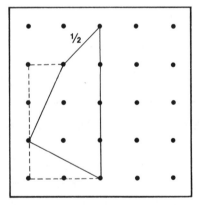

Chop off the small triangle at the top:

$$7 - \frac{1}{2} = 6\frac{1}{2}$$

Step 4

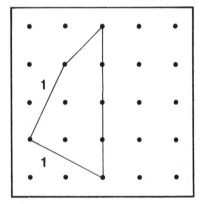

Chop off the two right triangles, each with area of 1 square unit.

$$6\frac{1}{2} - 2 = 4\frac{1}{2}$$

The area of the original polygon is $4\frac{1}{2}$ square units

Area of Polygons

Use Chris' chop strategy to find the area of each polygon. Show the rectangle you use in each case to surround the original figure.

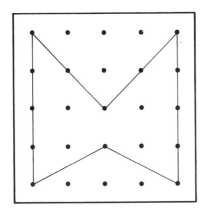

1. Area = _____

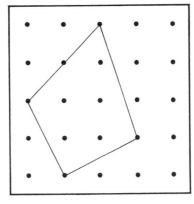

2. Area = _____

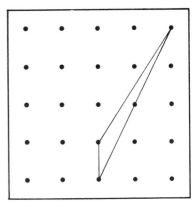

3. Area = _____

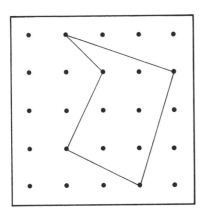

4. Area = _____

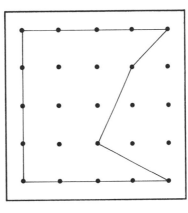

5. Area = _____

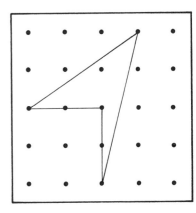

6. Area = _____

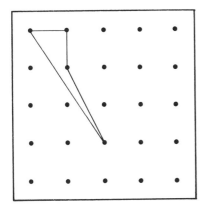

7. Area = _____

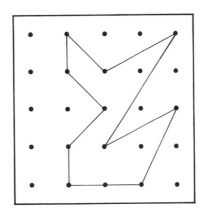

8. Area = _____

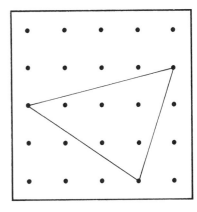

9. Area = _____

Area of Familiar Figures

Find the area of each figure. Let 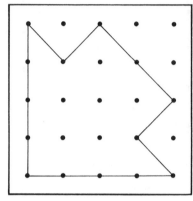 equal 1 square unit of area. Check your answers by doing them in more than one way.

1. Chair

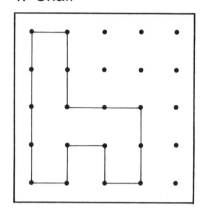

Area = _____

2. Shirt

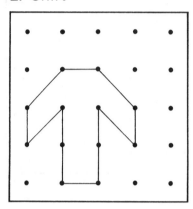

Area = _____

3. Big Arrow

Area = _____

4. Pipe

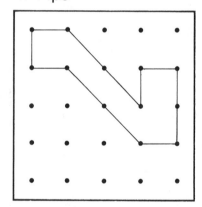

Area = _____

5. Bridge

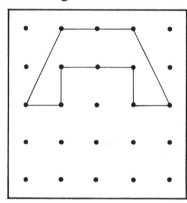

Area = _____

6. Shoe

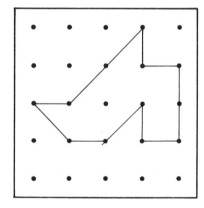

Area = _____

7. Crown

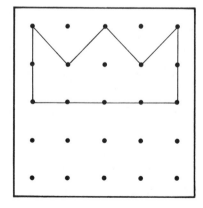

Area = _____

8. Knight's Helmet

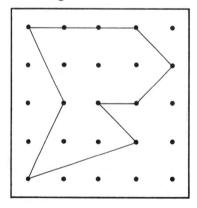

Area = _____

9. Your Choice: _____

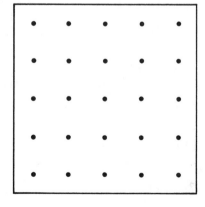

Area = _____

Area of Geoboard Numerals

Estimate and then find the area of each of these geoboard numerals. Omit the shaded regions.

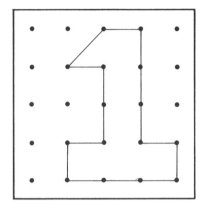

1. Estimated Area: _____
   Actual Area: _____

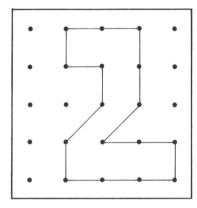

2. Estimated Area: _____
   Actual Area: _____

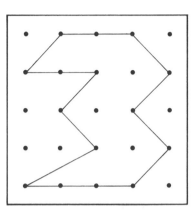

3. Estimated Area: _____
   Actual Area: _____

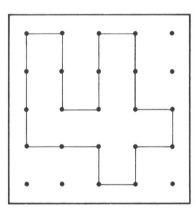

4. Estimated Area: _____
   Actual Area: _____

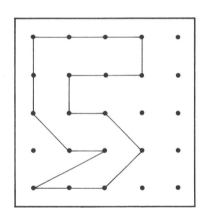

5. Estimated Area: _____
   Actual Area: _____

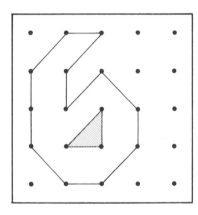

6. Estimated Area: _____
   Actual Area: _____

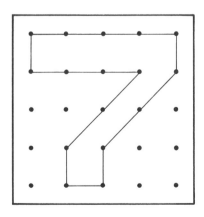

7. Estimated Area: _____
   Actual Area: _____

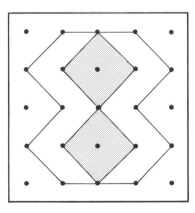

8. Estimated Area: _____
   Actual Area: _____

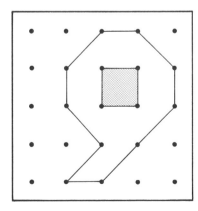

9. Estimated Area: _____
   Actual Area: _____

Alphabet Area, A-I

Estimate and then find the area of each of these geoboard letters.

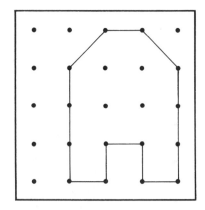

1. Estimated Area: _____
   Actual Area: _____

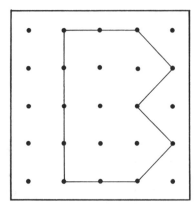

2. Estimated Area: _____
   Actual Area: _____

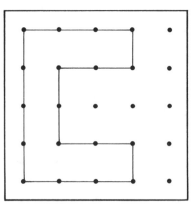

3. Estimated Area: _____
   Actual Area: _____

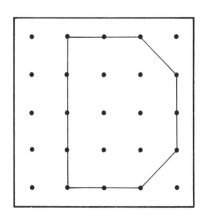

4. Estimated Area: _____
   Actual Area: _____

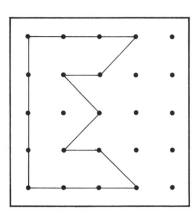

5. Estimated Area: _____
   Actual Area: _____

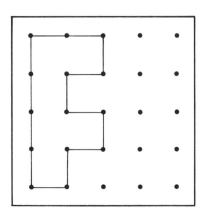

6. Estimated Area: _____
   Actual Area: _____

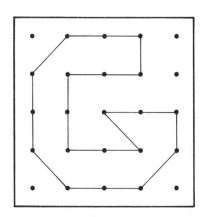

7. Estimated Area: _____
   Actual Area: _____

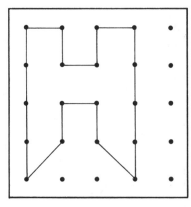

8. Estimated Area: _____
   Actual Area: _____

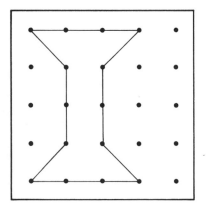

9. Estimated Area: _____
   Actual Area: _____

Alphabet Area, J-R

Estimate and then find the area of each of these geoboard letters.

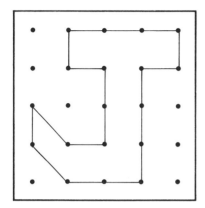

1. Estimated Area: _____
   Actual Area: _____

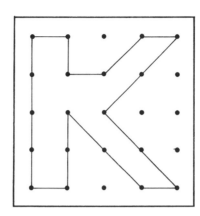

2. Estimated Area: _____
   Actual Area: _____

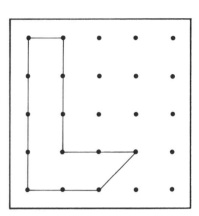

3. Estimated Area: _____
   Actual Area: _____

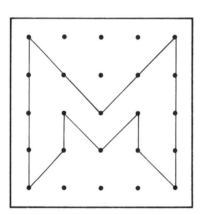

4. Estimated Area: _____
   Actual Area: _____

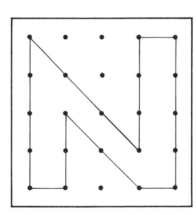

5. Estimated Area: _____
   Actual Area: _____

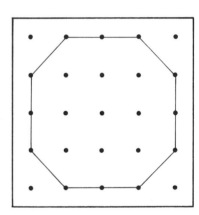

6. Estimated Area: _____
   Actual Area: _____

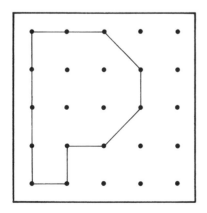

7. Estimated Area: _____
   Actual Area: _____

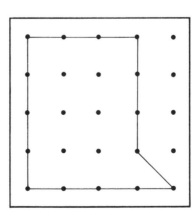

8. Estimated Area: _____
   Actual Area: _____

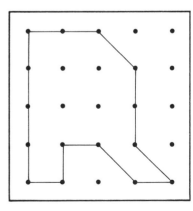

9. Estimated Area: _____
   Actual Area: _____

Alphabet Area, S-Z

Estimate and then find the area of each of these geoboard letters.

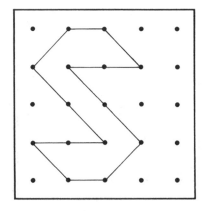

1. Estimated Area: _____
   Actual Area: _____

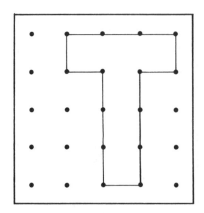

2. Estimated Area: _____
   Actual Area: _____

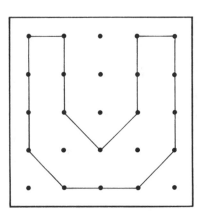

3. Estimated Area: _____
   Actual Area: _____

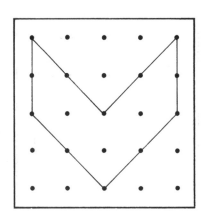

4. Estimated Area: _____
   Actual Area: _____

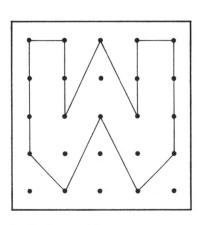

5. Estimated Area: _____
   Actual Area: _____

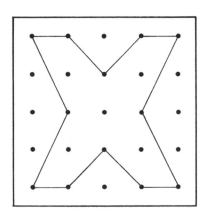

6. Estimated Area: _____
   Actual Area: _____

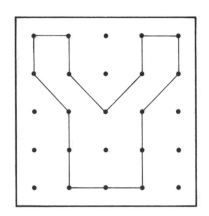

7. Estimated Area: _____
   Actual Area: _____

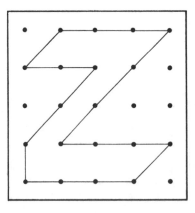

8. Estimated Area: _____
   Actual Area: _____

9. Write your first name in geoboard letters on a sheet of dot paper. What is the total area of your first name? _____

Area of Squares — Square Hunt

Try to construct a square with each of the given areas on your geoboard. Sketch your answers below. Three of these values are impossible. Compare your squares with your classmates.

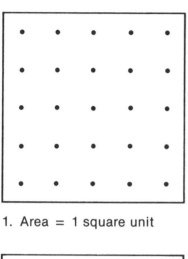

1. Area = 1 square unit

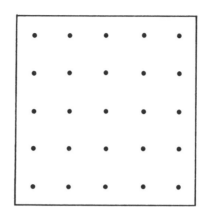

2. Area = 2 square units

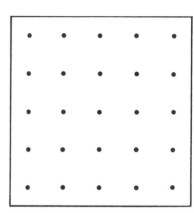

3. Area = 3 square units

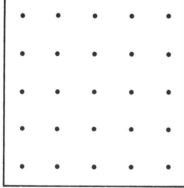

4. Area = 4 square units

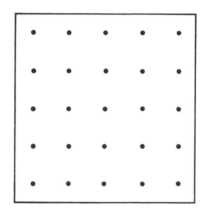

5. Area = 5 square units

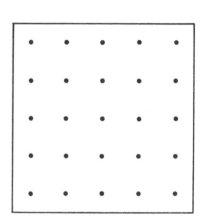

6. Area = 6 square units

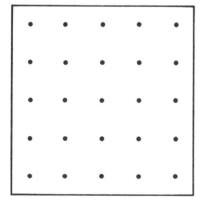

7. Area = 7 square units

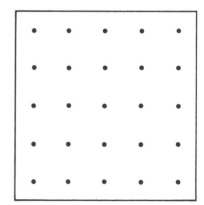

8. Area = 8 square units

9. Area = 9 square units

Area — Fish, Birds, Animals, and Objects

Find the area of each of these familiar objects made on more than one geoboard. Let ⬜ be 1 square unit of area. You may wish to work with a partner.

1. Shark

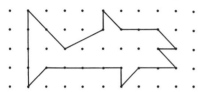

Area = _____

2. Kangaroo

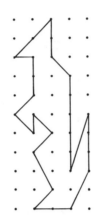

Area = _____

3. Dog

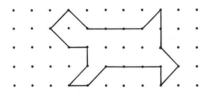

Area = _____

6. Candle

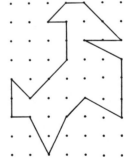

Area = _____

4. Sailboat

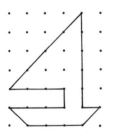

Area = _____

5. Whale

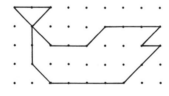

Area = _____

9. Goat

8. Vulture

7. Fox

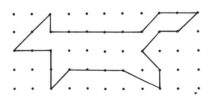

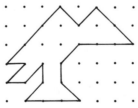

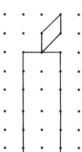

Area = _____   Area = _____   Area = _____

Area — Fish, Birds, Animals, and Objects

Find the area of each of these familiar objects made on more than one geoboard. Let ⬜ be 1 square unit of area. You may wish to work with a partner.

1. Cat

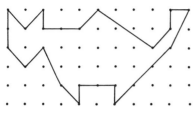

Area = _____

2. Rocket

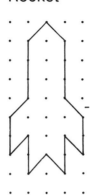

Area = _____

3. House

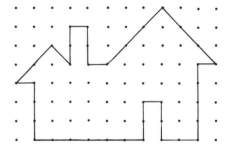

Area = _____

4. Microscope

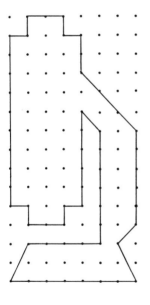

Area = _____

5. Sketch your own creation on the grid below and find its area.

Name of your creation

_____

Area = _____

Finding Dimensions Using Area

Use dot paper to help you find the missing dimensions for each shape.

1.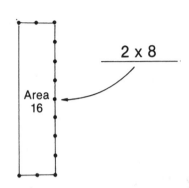

2.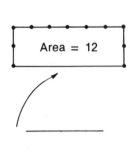

3.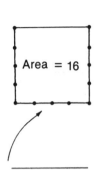

4.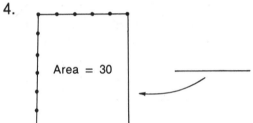

5.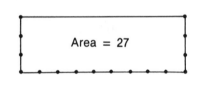

6.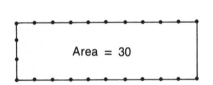

7.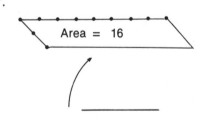

8.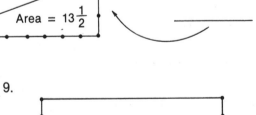

9.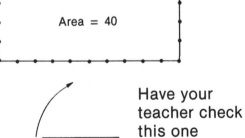

Have your teacher check this one

10.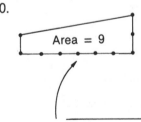

Lengths of Segments in Metric

Use your metric ruler to measure the length of each segment in millimeters (mm) and in centimeters (cm).

$$10 \text{ mm} = 1 \text{ cm}$$

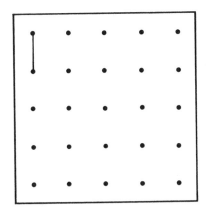

1. Length = _____mm
   Length = _____cm

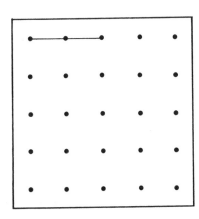

2. Length = _____ mm
   Length = _____ cm

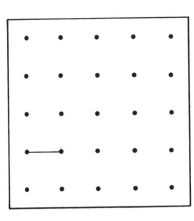

3. Length = _____mm
   Length = _____cm

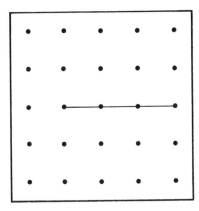

4. Length = _____mm
   Length = _____ cm

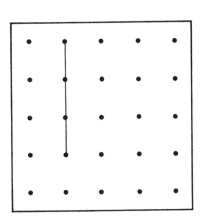

5. Length = _____mm
   Length = _____cm

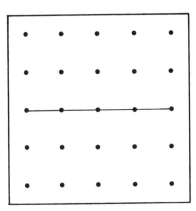

6. Length = _____mm
   Length = _____cm

When giving the length of any segment,
you must always record the unit of length used.

Measuring Lengths of Segments

Measure the length of each segment to the closest millimeter (mm) using your metric ruler.

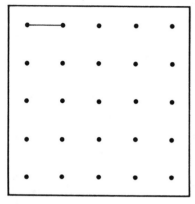

1. Length = _____mm

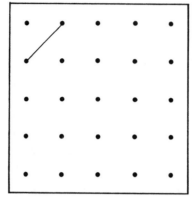

2. Length =_____ mm

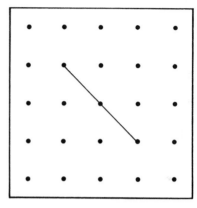

3. Length =_____mm

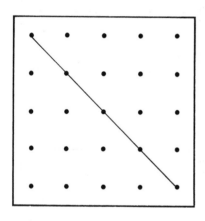

4. Length = _____mm

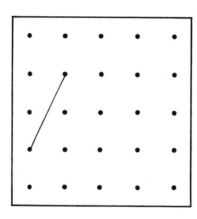

5. Length = _____mm

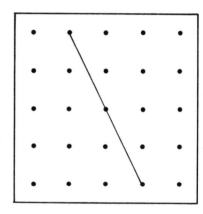

6. Length = _____mm

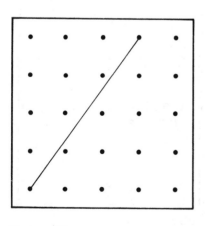

7. Length = _____mm

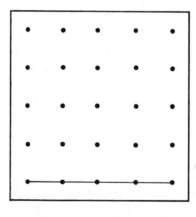

8. Length = _____mm

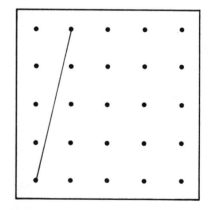

9. Length = _____mm

10. What is the length of the shortest segment that can be built on one geoboard? _____ mm
11. What is the length of the longest segment that can be built on one geoboard? _____ mm

47

Measuring Lengths of Segments

Measure the length of each segment to the closest millimeter (mm) using your metric ruler.

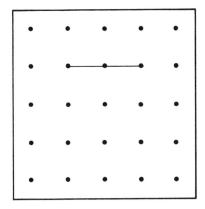

1. Length = _____mm

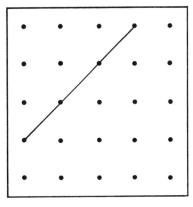

2. Length = _____ mm

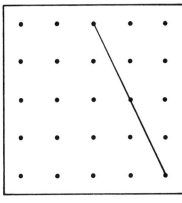

3. Length = _____mm

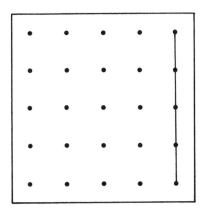

4. Length = _____mm

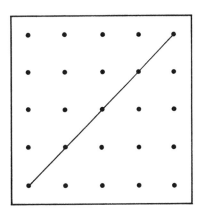

5. Length = _____mm

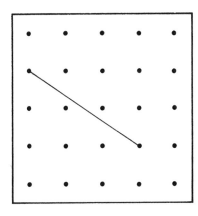

6. Length = _____mm

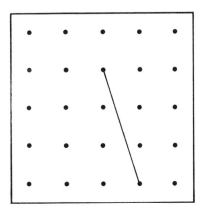

7. Length = _____mm

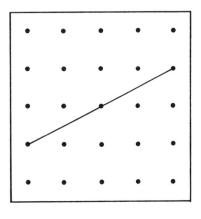

8. Length = _____mm

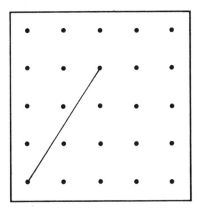

9. Length = _____mm

Perimeter of Polygons

Estimate and then measure the perimeter of each of these polygons to the closest centimeter (cm). The perimeter is the total distance around the figure.

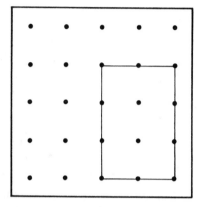

1. Estimated Perimeter:_____cm
   Actual   Perimeter:_____cm

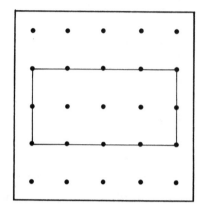

2. Estimated Perimeter:_____cm
   Actual   Perimeter:_____cm

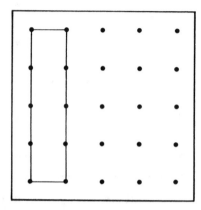

3. Estimated Perimeter:_____cm
   Actual   Perimeter:_____cm

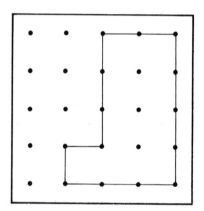

4. Estimated Perimeter:_____cm
   Actual   Perimeter:_____cm

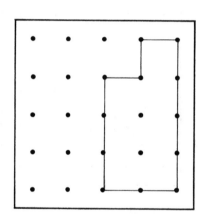

5. Estimated Perimeter:_____cm
   Actual   Perimeter:_____cm

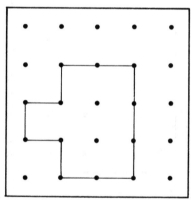

6. Estimated Perimeter:_____cm
   Actual   Perimeter:_____cm

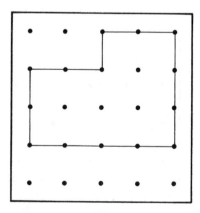

7. Estimated Perimeter:_____cm
   Actual   Perimeter:_____cm

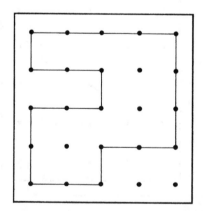

8. Estimated Perimeter:_____cm
   Actual   Perimeter:_____cm

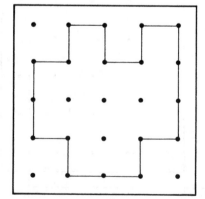

9. Estimated Perimeter:_____cm
   Actual   Perimeter:_____cm

DOT PAPER GEOMETRY © 1980 Cuisenaire Co. of America, Inc.

Perimeter of Polygons

Draw a polygon with each of the given perimeters. Compare your polygons with your classmates.

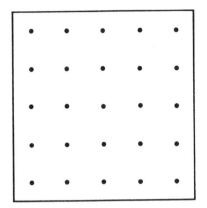

1. Perimeter = 10 units

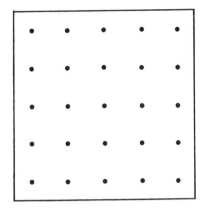

2. Perimeter = 12 units

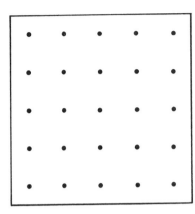

3. Perimeter = 8 units

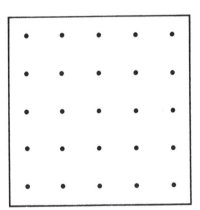

4. Perimeter = 20 units

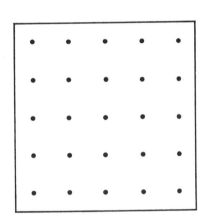

5. Perimeter = 16 units

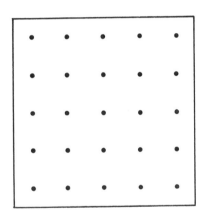

6. Perimeter = 18 units

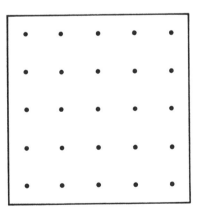

7. Perimeter = 11 units

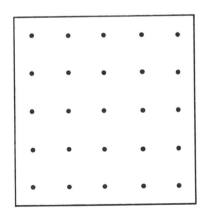

8. Perimeter = 14 units

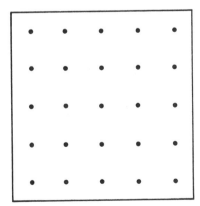

9. Perimeter = 17 units

Perimeter of Polygons

Estimate and then measure the perimeter of each of these polygons to the closest millimeter (mm).

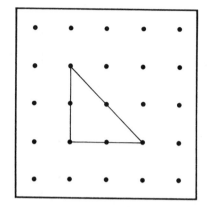

1. Estimated Perimeter:_____mm
   Measured Perimeter:_____mm

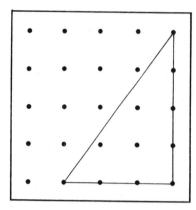

2. Estimated Perimeter:_____mm
   Measured Perimeter:_____mm

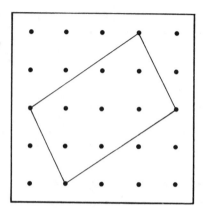

3. Estimated Perimeter:_____mm
   Measured Perimeter:_____mm

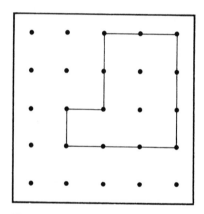

4. Estimated Perimeter:_____mm
   Measured Perimeter:_____mm

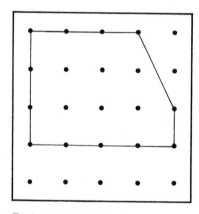

5. Estimated Perimeter:_____mm
   Measured Perimeter:_____mm

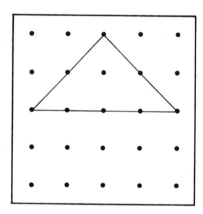

6. Estimated Perimeter:_____mm
   Measured Perimeter:_____mm

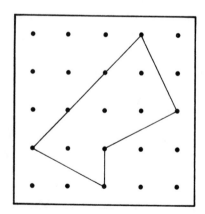

7. Estimated Perimeter:_____mm
   Measured Perimeter:_____mm

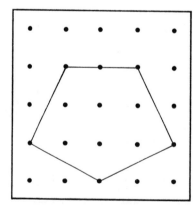

8. Estimated Perimeter:_____mm
   Measured Perimeter:_____mm

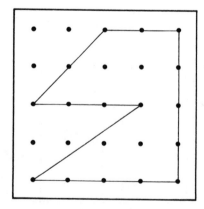

9. Estimated Perimeter:_____mm
   Measured Perimeter:_____mm

51

Perimeter of Geoboard Numerals

Estimate and then measure the perimeter of each of these geoboard numerals to the closest millimeter (mm).

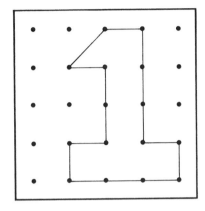
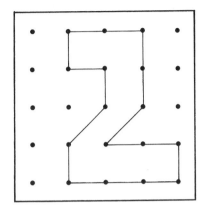
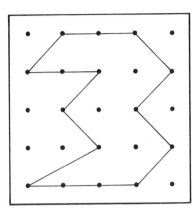

1. Estimated Perimeter:_____mm
   Measured Perimeter:_____mm

2. Estimated Perimeter:_____mm
   Measured Perimeter:_____mm

3. Estimated Perimeter:_____mm
   Measured Perimeter:_____mm

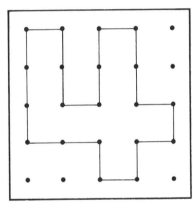
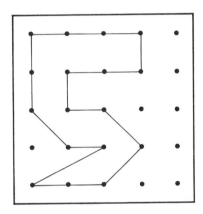
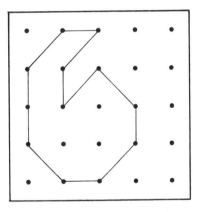

4. Estimated Perimeter:_____mm
   Measured Perimeter:_____mm

5. Estimated Perimeter:_____mm
   Measured Perimeter:_____mm

6. Estimated Perimeter:_____mm
   Measured Perimeter:_____mm

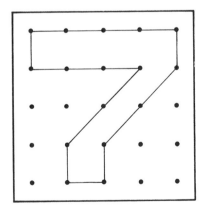
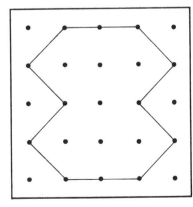
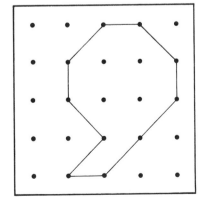

7. Estimated Perimeter:_____mm
   Measured Perimeter:_____mm

8. Estimated Perimeter:_____mm
   Measured Perimeter:_____mm

9. Estimated Perimeter:_____mm
   Measured Perimeter:_____mm

Alphabet Perimeter, A-I

Estimate and then measure the perimeter of each of these geoboard letters to the closest millimeter (mm).

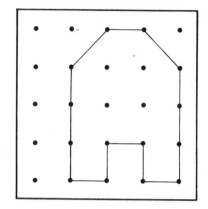

1. Estimated Perimeter:_____mm
   Measured Perimeter:_____mm

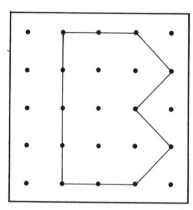

2. Estimated Perimeter:_____mm
   Measured Perimeter:_____mm

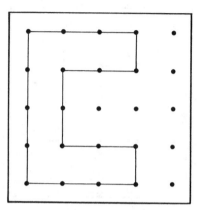

3. Estimated Perimeter:_____mm
   Measured Perimeter:_____mm

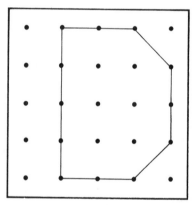

4. Estimated Perimeter:_____mm
   Measured Perimeter:_____mm

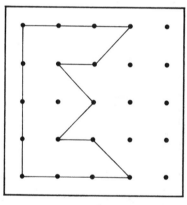

5. Estimated Perimeter:_____mm
   Measured Perimeter:_____mm

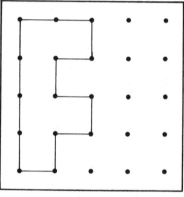

6. Estimated Perimeter:_____mm
   Measured Perimeter:_____mm

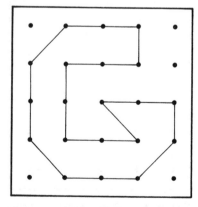

7. Estimated Perimeter:_____mm
   Measured Perimeter:_____mm

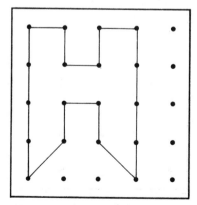

8. Estimated Perimeter:_____mm
   Measured Perimeter:_____mm

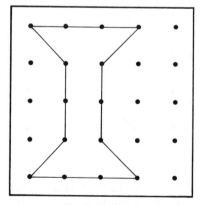

9. Estimated Perimeter:_____mm
   Measured Perimeter:_____mm

Alphabet Perimeter, J-R

Estimate and then measure the perimeter of each of these geoboard letters to the closest millimeter (mm).

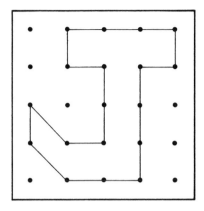

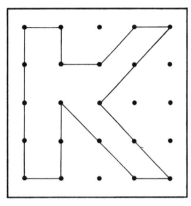

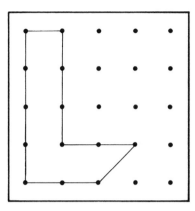

1. Estimated Perimeter:_____mm
   Measured Perimeter:_____mm

2. Estimated Perimeter:_____mm
   Measured Perimeter:_____mm

3. Estimated Perimeter:_____mm
   Measured Perimeter:_____mm

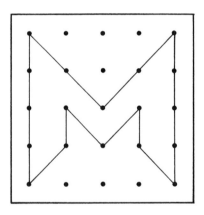

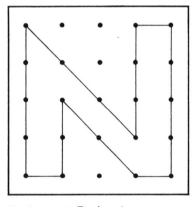

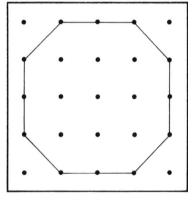

4. Estimated Perimeter:_____mm
   Measured Perimeter:_____mm

5. Estimated Perimeter:_____mm
   Measured Perimeter:_____mm

6. Estimated Perimeter:_____mm
   Measured Perimeter:_____mm

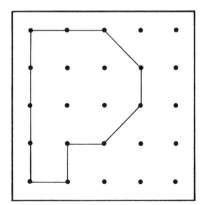

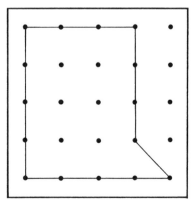

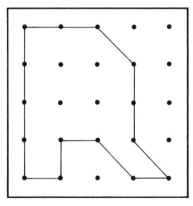

7. Estimated Perimeter:_____mm
   Measured Perimeter:_____mm

8. Estimated Perimeter:_____mm
   Measured Perimeter:_____mm

9. Estimated Perimeter:_____mm
   Measured Perimeter:_____mm

Alphabet Perimeter, S-Z

Estimate and then measure the perimeter of each of these geoboard letters to the closest millimeter (mm).

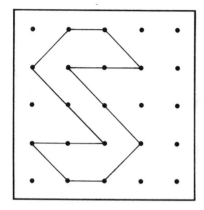 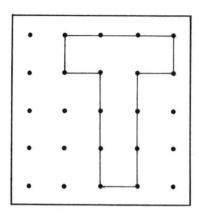 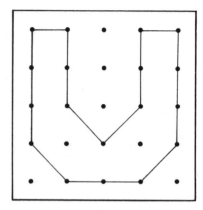

1. Estimated Perimeter:_____mm
   Measured Perimeter:_____mm

2. Estimated Perimeter:_____mm
   Measured Perimeter:_____mm

3. Estimated Perimeter:_____mm
   Measured Perimeter:_____mm

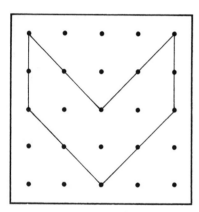 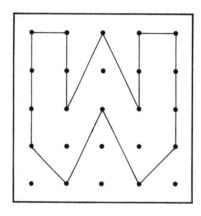 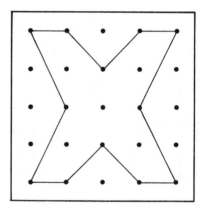

4. Estimated Perimeter:_____mm
   Measured Perimeter:_____mm

5. Estimated Perimeter:_____mm
   Measured Perimeter:_____mm

6. Estimated Perimeter:_____mm
   Measured Perimeter:_____mm

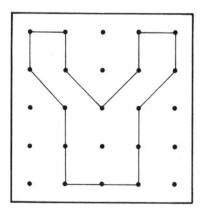 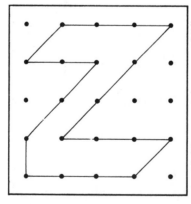

9. Write your initials in geoboard letters on a sheet of dot paper. What is the total perimeter of your initials to the closest centimeter (cm)?

_____ cm

7. Estimated Perimeter:_____mm.
   Measured Perimeter:_____mm.

8. Estimated Perimeter:_____mm
   Measured Perimeter:_____mm

Estimating and Measuring Angles

First estimate and then use your protractor to find the measure of each angle.

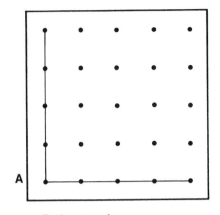

A

1. Estimate ∠A = _____
   Measurement ∠ A = _____

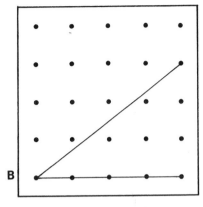

B

2. Estimate ∠ B = _____
   Measurement ∠ B = _____

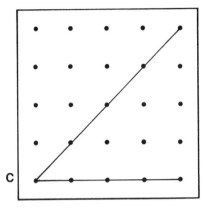

C

3. Estimate ∠ C = _____
   Measurement ∠ C = _____

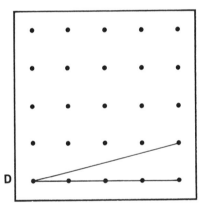

D

4. Estimate ∠D = _____
   Measurement ∠ D = _____

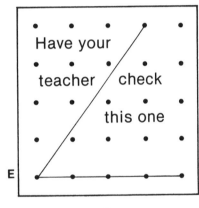

Have your teacher / check this one

E

5. Estimate ∠ E = _____
   Measurement ∠ E = _____

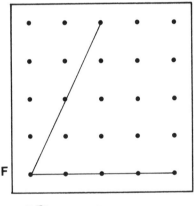

F

6. Estimate ∠ F = _____
   Measurement ∠ F = _____

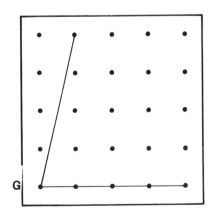

G

7. Estimate ∠G = _____
   Measurement ∠ G = _____

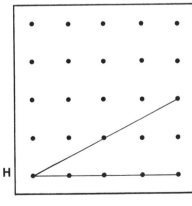

H

8. Estimate ∠ H = _____
   Measurement ∠ H = _____

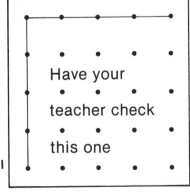

Have your teacher check this one

I

9. Estimate ∠ I = _____
   Measurement ∠ I = _____

Estimating and Measuring Angles

First estimate and then use your protractor to find the measure of each angle.

A

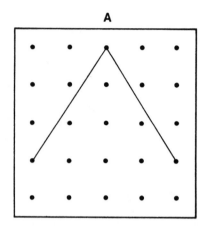

C

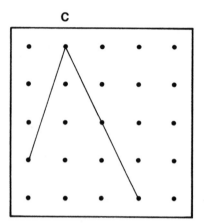

B

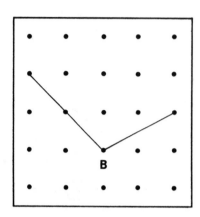

1. Estimate ∠A = _____
   Measurement ∠ A = _____

2. Estimate ∠ B = _____
   Measurement ∠ B = _____

3. Estimate ∠ C = _____
   Measurement ∠ C = _____

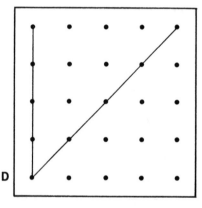

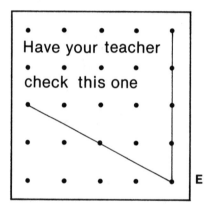

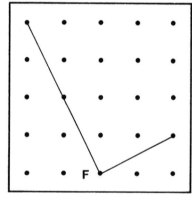

4. Estimate ∠D = _____
   Measurement ∠ D = _____

5. Estimate ∠ E = _____
   Measurement ∠ E = _____

6. Estimate ∠ F = _____
   Measurement ∠ F = _____

G

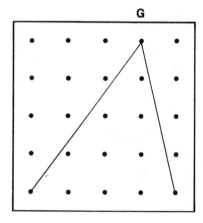

H

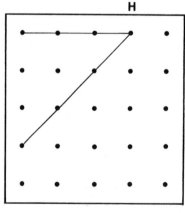

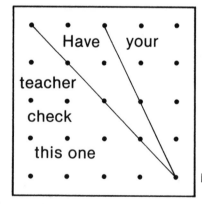

7. Estimate ∠G = _____
   Measurement ∠ G = _____

8. Estimate ∠ H = _____
   Measurement ∠ H = _____

9. Estimate ∠ I = _____
   Measurement ∠ I = _____

Measuring Angles of a Triangle

Use your protractor to measure each angle. Record each answer below. Then find the sum of the angles. Look for a pattern in the sums.

1.

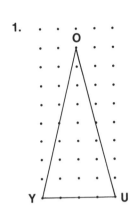

Measure ∠ Y = _____
Measure ∠ O = _____
Measure ∠ U = _____
_____

Sum of the angles = _____

2.

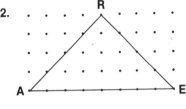

Measure ∠ A = _____
Measure ∠ R = _____
Measure ∠ E = _____
_____

Sum of the angles = _____

3.

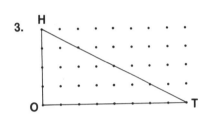

Measure ∠ H = _____
Measure ∠ O = _____
Measure ∠ T = _____
_____

Sum of the angles = _____

4.

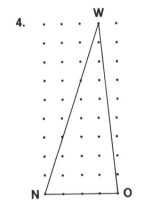

Measure ∠ N = _____
Measure ∠ O = _____
Measure ∠ W = _____
_____

Sum of the angles = _____

5.

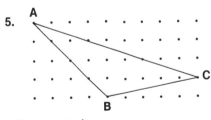

Measure ∠ A = _____
Measure ∠ B = _____
Measure ∠ C = _____
_____

Sum of the angles = _____

6. What pattern do you notice? _____

Do you think the pattern will hold for all triangles? _____

The Angle Sum of a Triangle

Use your protractor to measure each angle. Record each answer below. Then find the sum of the angles. Does your pattern hold for these triangles?

1.

2.

Measure ∠ E = _____
Measure ∠ D = _____
Measure ∠ F = _____

_____

Sum of the angles = _____

Measure ∠ G = _____
Measure ∠ H = _____
Measure ∠ J = _____

_____

Sum of the angles = _____

Challenge

Use the pattern you discovered to help find the measure of each angle labeled with a question mark. Check your answers using your protractor.

3.
65°   65°
?
? _____

4.
90°
?
45°
_____

5.
20°
45°
?
_____

6.
? _____
55°   55°

7.
25°
90°   ?
_____

8.
?
105°
45°
_____

DOT PAPER GEOMETRY © 1980 Cuisenaire Co. of America, Inc.

Measuring Angles of a Quadrilateral

Use your protractor to measure each angle. Record each answer below. Then find the sum of the angles. Look for a pattern in the sums.

1.

Measure ∠ A = _____
Measure ∠ B = _____
Measure ∠ C = _____
Measure ∠ D = _____

Sum of the angles = _____

2.

Measure ∠ E = _____
Measure ∠ F = _____
Measure ∠ G = _____
Measure ∠ H = _____

Sum of the angles = _____

3.

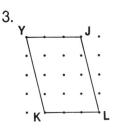

Measure ∠ Y = _____
Measure ∠ J = _____
Measure ∠ K = _____
Measure ∠ L = _____

Sum of the angles = _____

4.

Measure ∠ M = _____
Measure ∠ N = _____
Measure ∠ O = _____
Measure ∠ P = _____

Sum of the angles = _____

5.

Measure ∠ Q = _____
Measure ∠ R = _____
Measure ∠ S = _____
Measure ∠ T = _____

Sum of the angles = _____

6.

Measure ∠ U = _____
Measure ∠ V = _____
Measure ∠ W = _____
Measure ∠ X = _____

Sum of the angles = _____

7. What pattern do you notice? _____

   Do you think the pattern will hold for all quadrilaterals? _____

Measuring Angles of Polygons

1. Use your protractor to measure each angle of the following polygons. Find the sum of the angles for each one.

A.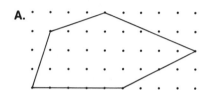

Sum of the angles = _____

B.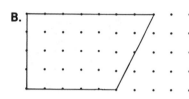

Sum of the angles = _____

C.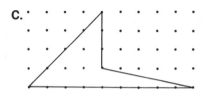

Sum of the angles = _____

D.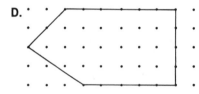

Sum of the angles = _____

E.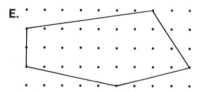

Sum of the angles = _____

F.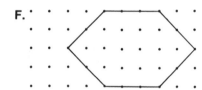

Sum of the angles = _____

G.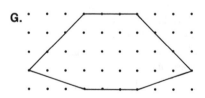

Sum of the angles = _____

H.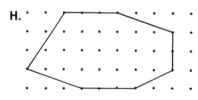

Sum of the angles = _____

I.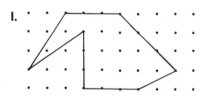

Sum of the angles = _____

Measuring Angles of Polygons (continued)

1. Record the angle sums from the previous page in the chart below according to the number of sides of each polygon.

How many sides?	What is the Sum of the Angles?
3	180°
4	
5	
6	
7	

What patterns do you notice?

_____

_____

_____

2. Predict the angle sum for this 8-sided figure. Check your prediction by measuring.

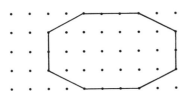

Predicted sum of the angles _____
Measured sum of the angles _____

3. Now predict the sum of the angles for a polygon with 9, 10, 11, 12, 100, 1001, or N sides. Write your predictions in this chart.

Number of sides?	Predicted Angle Sum
9	
10	
11	
12	
100	
1001	
N (any number)	

Diagonals of a Polygon

1. Construct all the diagonals for each polygon. Record your answers in the table below.

A.

B.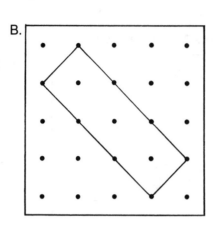

C.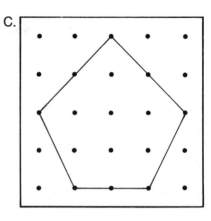

D.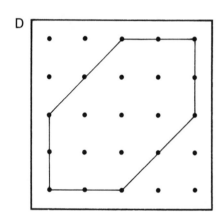

E.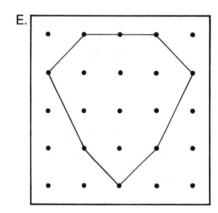

F.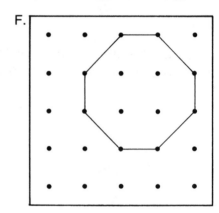

	A	B	C	D	E	F
Number of sides	3	4	5	6	7	8
Number of diagonals						

2. What patterns do you notice? _____

3. Predict how many diagonals a nine-sided polygon will have. _____
   Make one on dot paper and check your prediction.

4. Predict how many diagonals a ten-sided polygon will have. _____
   Make one on dot paper and check your prediction.

5. Try to generalize for an n-sided polygon. _____

DOT PAPER GEOMETRY © 1980 Cuisenaire Co. of America, Inc.

Dividing a Geoboard into Regions

Jane was explaining the maximum number of regions that can be found on a geoboard using 1, 2, 3, 4, 5, 6, 7, 8, or 9 segments. With 1 segment, she found 2 regions. With 2 segments, she found 4 regions. With 3 segments, she found 7 regions.

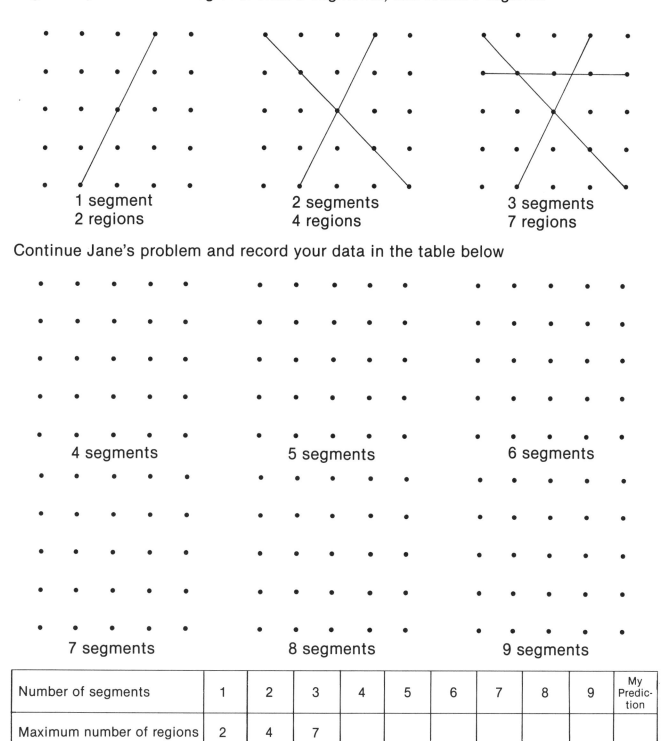

1 segment
2 regions

2 segments
4 regions

3 segments
7 regions

Continue Jane's problem and record your data in the table below

4 segments

5 segments

6 segments

7 segments

8 segments

9 segments

Number of segments	1	2	3	4	5	6	7	8	9	My Prediction
Maximum number of regions	2	4	7							

What patterns do you notice? _____

_____

Metric Geoboard Template — Master

Geoboard Grids — Master

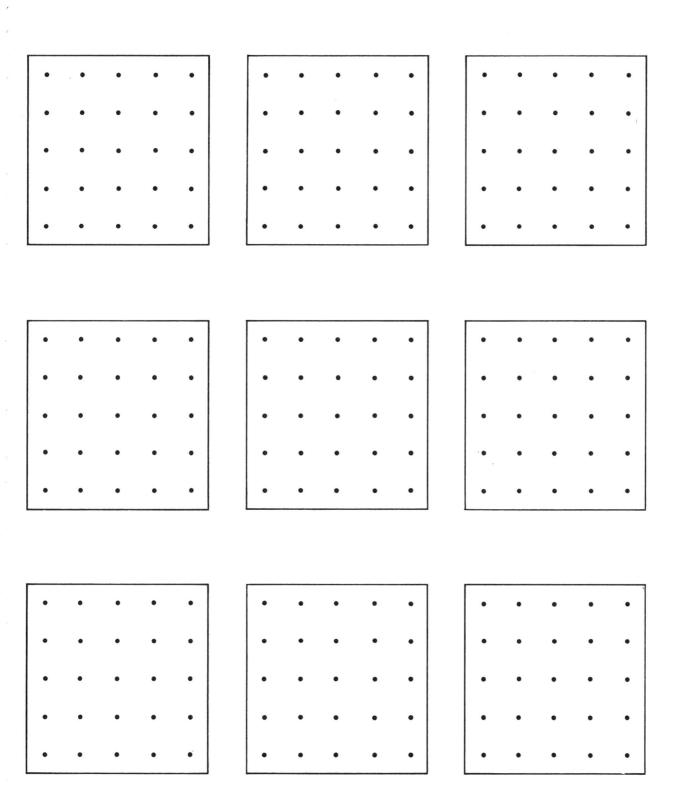

Geoboard Dot Paper — Master

Quiz: Area of Polygons

Name: _____ Date: _____

Circle the letter of the correct answer for each problem. Let ▢ be 1 square unit of area.

1. What is the area of this shape?

   A. 1 square unit   B. 4 square units   C. 10 square units   D. None of these is correct.

2. Which one of the following shapes has an area of 2½ square units?

   A.   B.   C.   D.   E. None of these.

3. Which one of the following shapes does not have an area of 3½ square units?

   A.   B.   C.   D.   E. All of these have an area of 3½ square units.

4. What is the area of the shaded region shown below?

   A. 4 square units   B. 6 square units   C. 10 square units   D. 16 square units   E. None of these.

5. What is the area of this shape?

   A. 6 square units   B. 6½ square units   C. 7 square units   D. 7½ square units   E. 8 square units

6. Which of the following is the correct sketch of a square whose area is 5 square units?

   A.   B.   C.   D.   E. None of these

Quiz: Area of Polygons (continued)

Name: _____ Date: _____

Circle the letter of the correct answer for each problem. Let ▢ be 1 square unit of area.

7. What is the area of this triangle?

   A. 7 square units   B. 8 square units   C. 12 square units   D. Impossible to determine   E. None of these
   is correct

8. Jim said the area of the rectangle shown below is 5½ square units. Mike said it is 6 square units.
   Who is correct?

   A. Jim   B. Mike   C. Jim & Mike are both correct.   D. Jim & Mike are both wrong.

9. Pam and Kelly placed 4 ordinary geoboards together like this:

   What is the area of the largest square which they can
   construct on their giant geoboard?

   A. 64 square units   B. 81 square units   C. 100 square units

   D. 16 square units   E. Impossible to determine.

10. Ken placed 6 ordinary geoboards in a line like this:

   What is the area of the largest rectangle he can build on his new creation?
   A. 56 square units   B. 70 square units   C. 80 square units   D. 116 square units   E. 150 square
   units

DOT PAPER GEOMETRY © 1980 Cuisenaire Co. of America, Inc.

Quiz: Perimeter of Polygons

Name:_____ Date:_____

Circle the letter of the correct answer for each problem.

1. Select the triangle with the <u>smallest</u> perimeter.

2. Select the square with the <u>largest</u> perimeter.

3. Select the rectangle with the <u>largest</u> perimeter.

A.

A.

A.

B.

B.

B.

C.

C.

C.

D.

D.

D.

Quiz: Perimeter of Polygons (continued)

Name: _____ Date: _____

Circle the letter of the correct answer for each problem.

4. Select the quadrilateral with the largest perimeter

  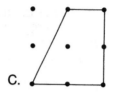 

5. Select the polygon with the largest perimeter.

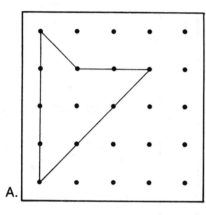 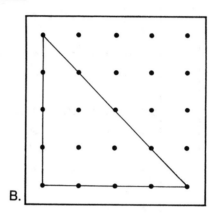

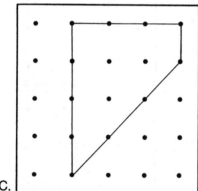 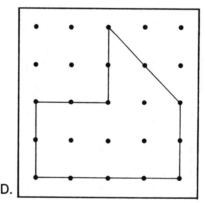

6. What is the perimeter of this rectangle?

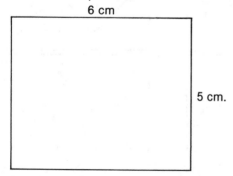

A. 11 cm. B. 22 cm. C. 30 cm.
D. 65 cm. E. None of these

7. What is the perimeter of this square?

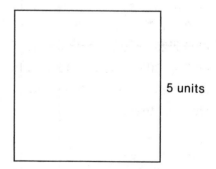

A. 5 units   B. 20 units   C. 25 units
D. 5 by 5 units   E. None of these.

Name: _____ Date: _____

Circle the letter of the correct answer for each problem.

8. Here are 4 squares. What is the closest value for the total perimeter of the 4 squares?

   A. 16 cm.
   B. 48 cm.
   C. 120 cm.
   D. 144 cm.
   E. None of these

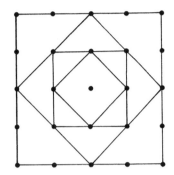

9. Jim placed 3 geoboards side by side like this:

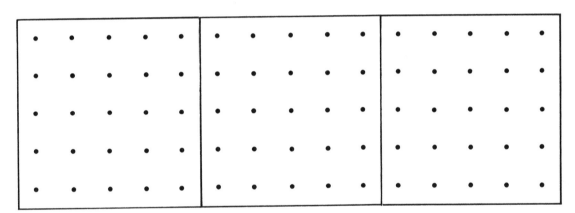

What is the perimeter of the largest rectangle which can be constructed on his creation?

   A. 48 cm.   C. 98 cm.   E. None of these.
   B. 56 cm.   D. 108 cm.

10. Mr. Wells wanted to put a fence around his yard. How much fence does he need to buy?

   A. (200 x 150) units
   B. (200 x 200 x 150 x 150) units
   C. (200 + 200 + 150 + 150) units
   D. (200 x 200 + 150 x 150) units
   E. None of these.

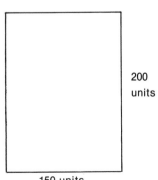

200 units

150 units

Selected Answers and Commentary

Pages 1-18: Introduction to Polygons and their Properties

Page 1:

	No. of nails	No. of sides
B	12	7
C	18	8
D	12	6
E	16	9
F	Answers will vary	
G	Answers will vary	
H	18	12
I	14	8

Page 2:

	No. of nails	No. of sides
2)	16	11
3)	15	11
4)	20	14
5)	17	12
6)	Answers will vary	
7)	Answers will vary	
8)	12	10
9)	10	9

Page 3: Only 3 answers are required for each problem, but all possibilities are shown.

1) 6   2) 8   3) 6   4) 3   5) 2   6) 6   7) 4   8) 7   9) 8

Page 4: Only 3 answers are required for each problem, but all possibilities are shown.

1) 7   2) 4   3) 5   4) 4   5) 7   6) 5   7) 4   8) 4   9) 6

Page 7: Only 1 possible answer is shown.

1)   2)   3)   4)

Page 8: Polygon positions on the boards will vary.

1)   2)   3)   4)

Page 11: 1) C 2) E 3) Triangle 4) Square 5) Answers will vary

Page 12:
1) P (1,0), (1,1), (1,2), (1,3), (1,4), (2,4), (2,2), (3,2), (3,3), (3,4)
2) A (0,0), (4,0), (1,2), (2,2), (3,2), (2,4)
3) L (1,0), (1,1), (1,2), (1,3), (1,4), (2,0), (3,0), (4,0)
4) 4 (3,0), (3,1), (3,2), (3,3), (3,4), (4,2), (2,2), (1,2), (1,3), (1,4)

5) 9 (2,0), (3,0), (4,1), (4,2), (3,2), (2,2), (1,2), (1,3), (2,4), (3,4), (4,3)

6) 6 (2,4), (1,4), (0,3), (0,2), (0,1), (1,0), (2,0), (3,1), (2,2), (1,2)

Page 13:
1) Isosceles
2) Isosceles
3) Isosceles
4) Scalene
5) Scalene
6) Isosceles

Page 14:
1) Obtuse
2) Right
3) Acute
4) Right
5) Right
6) Obtuse

Page 15:
1) Square
   Rectangle
   Parallelogram
   Quadrilateral
2) Parallelogram
   Quadrilaterial
3) Trapezoid
   Quadrilateral
4) Trapezoid
   Quadrilateral
5) Square
   Rectangle
   Parallelogram
   Quadrilateral
6) Trapezoid
   Quadrilateral
7) Square
   Rectangle
   Parallelogram
   Quadrilateral
8) Rectangle
   Parallelogram
   Quadrilateral
9) Rectangle
   Parallelogram
   Quadrilateral

Page 16:
1) Quadrilateral
2) Parallelogram
   Quadrilateral
3) Square
   Rectangle
   Parallelogram
   Quadrilateral
4) Parallelogram
   Quadrilateral
5) Quadrilateral
6) Quadrilateral
7) Quadrilateral
8) Trapezoid
   Quadrilateral
9) Quadrilateral

Page 17:
If additional practice is needed with this type of activity, you may wish to have students work in pairs making and copying polygons. Have one person make the initial figure and his/her partner copy it in a different position or orientation on a geoboard. Record both of the figures on dot paper.

Page 18:
Exercises 4, 5, and 6 provide an opportunity for students to create figures with line symmetry of their own. You may wish to have them copy some of their creations on large dot paper (see appendix) and then display their work on the classroom bulletin board. If additional practice is needed with this type of activity, you may wish to have students work in pairs to create and complete figures so that they have line symmetry.

Pages 19-30: Fractions, Decimals, and Percents

Page 19:
Sample Solutions

1)   2)   3)

Page 20:
Sample Solutions

1) 2) 3) 4) 5) 6) 7) 8) 9)

Page 21:
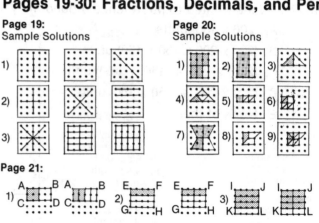
1)   2)   3)

Page 22:
Sample Solutions

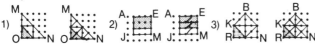

1) M O N   M O N   2) A. J...M E...M   A. J...M E...M   3) K R...N B   K R...N B

Page 26:

a) A...b) C...D   A...B C...D   c) A...B C...D   d) A...B C...D   e) A...B C...D

2) 3,3   5) 6,6   8) 2,2   11) 12,12
3) 2,2   6) 12,12 9) 4,4   12) 18,18
4) 4,4   7) 8,8   10) 2,2

Page 27:

1) $\frac{34}{100}$ .34 34%   2) $\frac{20}{100}$ .20 20%

Page 28:

1) $\frac{50}{100}$ .50 50%   3) $\frac{100}{100}$ 1.00 100%

2) $\frac{47}{100}$ .47 47%   4) $\frac{75}{100}$ .75 75%

Page 29:

1) $\frac{8}{25} \times \frac{4}{4} = \frac{32}{100} = 32\%$ 4) $\frac{24}{25} = \frac{96}{100} = 96\%$

2) $\frac{3}{25} = \frac{12}{100} = 12\%$ 5) $\frac{19}{25} = \frac{76}{100} = 76\%$

3) $\frac{11}{25} = \frac{44}{100} = 44\%$ 6) $\frac{2}{25} = \frac{8}{100} = 8\%$

Page 30:

1) $\frac{5}{25}$ $\frac{20}{100}$ 3) $\frac{12}{25}$ $\frac{48}{100}$ 5) $\frac{10}{25}$ $\frac{40}{100}$

2) $\frac{15}{25}$ $\frac{60}{100}$ 4) $\frac{22}{25}$ $\frac{88}{100}$ 6) $\frac{3}{25}$ $\frac{12}{100}$

Pages 31-45 : Area of Polygons

Page 31:
1) 6 sq. units
2) 12 sq. units
3) 8 sq. units
4) 9 sq. units X
5) 12 sq. units
6) 16 sq. units X

Page 32:
1) 24 sq. units
2) 14 sq. units
3) 24 sq. units
4) 36 sq. units

Page 33:
1) 2 sq. units
2) 2 sq. units
3) 4 sq. units
4) 1 sq. units
5) 4½ sq. units
6) ½ sq. unit

Page 34:
1) 4 sq. units
2) 4½ sq. units
3) 4 sq. units

Page 36
1) 10 sq. units
2) 6½ sq. units
3) 1 sq. units
4) 6½ sq. units
5) 11½ sq. units
6) 3 sq. units
7) 1 sq. unit
8) 6½ sq. units
9) 5½ sq. units

Page 37:
1) 7 sq. units
2) 5 sq. units
3) 12 sq. units
4) 5 sq. units
5) 4 sq. units
6) 6 sq. units
7) 6 sq. units
8) 9 sq. units
9) Answers will vary

Page 38:
1) 6½ sq. units
2) 7 sq. units
3) 9½ sq. units
4) 9 sq. units

5) 7½ sq. units
6) 7½ sq. units
7) 7 sq. units
8) 8 sq. units
9) 7 sq. units

Page 39:
1) 10 sq. units
2) 10 sq. units
3) 8 sq. units
4) 11 sq. units
5) 8 sq. units
6) 6 sq. units
7) 9 sq. units
8) 8 sq. units
9) 6 sq. units

Page 40:
1) 8 sq. units
2) 8½ sq. units
3) 5½ sq. units
4) 8 sq. units
5) 9½ sq. units
6) 14 sq. units
7) 9 sq. units
8) 12½ sq. units
9) 10½ sq. units

Page 41:
1) 6 sq. units
2) 6 sq. units
3) 10 sq. units
4) 8 sq. units
5) 11 sq. units
6) 10 sq. units
7) 8 sq. units
8) 8½ sq. units
9) You may wish to use the results of this execise as a bulletin board display

Page 42:
Sample solutions are shown below. However, it is recommended that teachers withold the answers from students on these exercises. Several of the problems are really quite challenging. As problems are posed and solved, you may wish to have students record their answers on a copy of one of the large record sheets provided in the appendix. This will facilitate displaying squares that are being "hunted" and those that have been "found". Many children will also enjoy extending this exercise by finding squares whose vertices are not at a nail or dot.

1) □   2) ◇   3) Impossible   4) ⊡   5) ◇

6) Impossible   7) Impossible   8) ◈   9) ⊡

Page 43:
1) 15½ sq. units
2) 18½ sq. units
3) 13 sq. units
4) 13 sq. units
5) 16 sq. units
6) 20 sq. units
7) 16½ sq. units
8) 14 sq. units
9) 21½ sq. units

Page 44:
1) 24 sq. units
2) 18 sq. units
3) 46½ sq. units
4) 67½ sq. units

Page 45:
1) 2x8
2) 2x6
3) 4x4
4) 5x6
5) 3x9
6) 3x10
7) 2x8
8) 3x9
9) 4x10
10) 2x6

Pages 46-55: Metric Length and Perimeter of Polygons

Unlike the other sections of this book, the answers to worksheets on length and perimeter will vary according to the size of the grid which is used. In order to provide assistance to teachers who may wish to provide experiences for their students by using a geoboard and dot paper, two sets of answers are provided in this section. The first set of answers for each page presumes the use of a 5x5 geoboard where nails are placed at 4 cm intervals. The second set of answers assumes the use of cm dot paper.

Page 46: Geoboard answers (4 cm between nails).

1) 40 mm, 4 cm
2) 80 mm, 8 cm
3) 40 mm, 4 cm
4) 120 mm, 12 cm
5) 120 mm, 12 cm
6) 160 mm, 16 cm

Page 46: Dot paper answers (1 cm between dots)

1) 10 mm, 1 cm
2) 20 mm, 2 cm
3) 10 mm, 1 cm
4) 30 mm, 3 cm
5) 30 mm, 3 cm
6) 40 mm, 4 cm

Page 47: Geoboard answers (4 cm between nails).

1) 40 mm
2) 56 mm
3) 112 mm
4) 224 mm
5) 88 mm
6) 179 mm
7) 200 mm
8) 160 mm
9) 164 mm
10) 40 mm
11) 226 mm

Page 47: Dot paper answers (1 cm between dots).

1) 10 mm
2) 14 mm
3) 28 mm
4) 56 mm
5) 22 mm
6) 45 mm
7) 50 mm
8) 40 mm
9) 41 mm
10) 10 mm
11) 56 mm

Page 48: Geoboard answers (4 cm between nails).

1) 80 mm
2) 168 mm
3) 176 mm
4) 160 mm
5) 224 mm
6) 144 mm
7) 128 mm
8) 176 mm
9) 144 mm

Page 48: Dot paper answers (1 cm between dots).

1) 20 mm
2) 42 mm
3) 44 mm
4) 40 mm
5) 56 mm
6) 36 mm
7) 32 mm
8) 44 mm
9) 36 mm

Page 49: Geoboard answers (4 cm between nails).

1) 40 cm
2) 48 cm
3) 40 cm
4) 56 cm
5) 48 cm
6) 48 cm
7) 56 cm
8) 80 cm
9) 72 cm

Page 49: Dot paper answers (1 cm between dots)

1) 10 cm
2) 12 cm
3) 10 cm
4) 14 cm
5) 12 cm
6) 12 cm
7) 14 cm
8) 20 cm
9) 18 cm

Page 50: Geoboard and dot paper.
Sample solutions are shown. Problems 7 and 9 are impossible if the vertices of the polygon must be at a dot or nail.

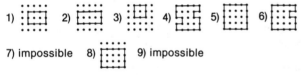

7) impossible 8) 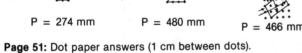 9) impossible

Page 51: Geoboard answers (4 cm between nails). Answers are approximate.

1) P = 274 mm
2) P = 480 mm
3) P = 466 mm

Page 51: Dot paper answers (1 cm between dots).

1) P = 68 mm
2) P = 120 mm
3) P = 116 mm

Page 52 Geoboard answers (4 cm between nails). Answers are approximate.

1) 616 mm
2) 672 mm
3) 760 mm
4) 800 mm
5) 736 mm
6) 576 mm
7) 704 mm
8) 608 mm
9) 496 mm

Page 52: Dot paper answers. (1 cm between dots).

1) 154 mm
2) 168 mm
3) 190 mm
4) 200 mm
5) 184 mm
6) 144 mm
7) 176 mm
8) 152 mm
9) 124 mm

Page 53: Geoboard answers (4 cm between nails). Answers are approximate.

1) 592 mm
2) 544 mm
3) 720 mm
4) 512 mm
5) 704 mm
6) 560 mm
7) 864 mm
8) 752 mm
9) 624 mm

Page 53: Dot paper answers (1 cm between dots). Answers are approximate.

1) 148 mm
2) 136 mm
3) 180 mm
4) 128 mm
5) 176 mm
6) 140 mm
7) 216 mm
8) 188 mm
9) 156 mm

Page 54: Geoboard answers (4 cm between nails). Answers are approximate.

1) 752 mm
2) 872 mm
3) 536 mm
4) 848 mm
5) 920 mm
6) 544 mm
7) 512 mm
8) 616 mm
9) 648 mm

Page 54: Dot paper answers (1 cm between dots). Answers are approximate.

1) 188 mm
2) 218 mm
3) 134 mm
4) 212 mm
5) 230 mm
6) 136 mm
7) 128 mm
8) 154 mm
9) 162 mm

Page 55: Geoboard answers (4 cm between nails). Answers are approximate.

1) 688 mm
2) 560 mm
3) 784 mm
4) 608 mm
5) 944 mm
6) 736 mm
7) 704 mm
8) 872 mm
9) Answers will vary

Page 55: Dot paper answers (1 cm between dots).

1) 172 mm
2) 140 mm
3) 196 mm
4) 152 mm
5) 236 mm
6) 184 mm
7) 176 mm
8) 218 mm
9) Answers will vary

Page 56-62 Teacher Note;

If you are using a dot paper approach, the angles shown on pages 56-62 may be difficult for students to measure directly on each worksheet. To make it easier to measure each angle, have students extend the lengths of the sides of the angles, or allow them to construct the same angles on the larger dot paper sheet in the appendix.

Page 56: Answers are approximations.

1) 90°
2) 37°
3) 45°
4) 14°
5) 53°
6) 65°
7) 77°
8) 27°
9) 90°

Page 57: Answers are approximations.

1) 67°
2) 108°
3) 45°
4) 45°
5) 63°
6) 90°
7) 60°
8) 45°
9) 19°

Page 58:

1) Y = 75°
   O = 30°
   U = 75°
   ‾‾‾‾
   180°

2) A = 45°
   R = 90°
   E = 45°
   ‾‾‾‾
   180°

3) H = 60°
   O = 90°
   T = 30°
   ‾‾‾‾
   180°

4) N = 70°
   O = 85°
   W = 25°
   ‾‾‾‾
   180°

5) A = 25°
   B = 125°
   C = 30°
   ‾‾‾‾
   180°

6) The sum of the 3 angles of each triangle is 180. Yes.

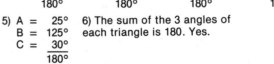
DOT PAPER GEOMETRY © 1980 Cuisenaire Co. of America, Inc.